McGraw-Hill Networking and Telecommunications

Build Your Own

TRULOVE • *Build Your Own Wireless LAN* (with projects)

Crash Course

LOUIS • *Broadband Crash Course*
VACCA • *i-Mode Crash Course*
LOUIS • *M-Commerce Crash Course*
SHEPARD • *Telecom Convergence*, 2/e
SHEPARD • *Telecom Crash Course*
LOUIS • *Telecom Management Crash Course*
BEDELL • *Wireless Crash Course*
KIKTA/FISHER/COURTNEY • *Wireless Internet Crash Course*

Demystified

HARTE/LEVINE/KIKTA • *3G Wireless Demystified*
LaROCCA • *802.11 Demystified*
MULLER • *Bluetooth Demystified*
EVANS • *CEBus Demystified*
BAYER • *Computer Telephony Demystified*
HERSHEY • *Cryptography Demystified*
TAYLOR • *DVD Demystified*
HOFFMAN • *GPRS Demystified*
SYMES • *MPEG-4 Demystified*
CAMARILLO • *SIP Demystified*
SHEPARD • *SONET / SDH Demystified*
TOPIC • *Streaming Media Demystified*
SYMES • *Video Compression Demystified*
SHEPARD • *Videoconferencing Demystified*
BHOLA • *Wireless LANs Demystified*

Developer Guides

VACCA • *i-Mode Crash Course*
GUTHERY/CRONIN • *Mobile Application Development with SMS*
RICHARD • *Service and Device Discovery: Protocols and Programming*

Professional Telecom

SMITH/COLLINS • *3G Wireless Networks*
BATES • *Broadband Telecom Handbook*, 2/e
COLLINS • *Carrier Grade Voice over IP*
HARTE • *Delivering xDSL*
HELD • *Deploying Optical Networking Components*
MINOLI • *Ethernet-Based Metro Area Networks*
BENNER • *Fibre Channel for SANs*
BATES • *GPRS*

MINOLI • *Hotspot Networks: WiFi for Public Access Locations*
LEE • *Lee's Essentials of Wireless*
BATES • *Optical Switching and Networking Handbook*
WETTEROTH • *OSI Reference Model for Telecommunications*
SULKIN • *PBX Systems for IP Telephony*
RUSSELL • *Signaling System #7*, 4/e
SAPERIA • *SNMP on the Edge: Building Service Management*
MINOLI • *SONET-Based Metro Area Networks*
NAGAR • *Telecom Service Rollouts*
LOUIS • *Telecommunications Internetworking*
RUSSELL • *Telecommunications Protocols*, 2/e
MINOLI • *Voice over MPLS*
KARIM/SARRAF • *W-CDMA and cdma2000 for 3G Mobile Networks*
BATES • *Wireless Broadband Handbook*
FAIGEN • *Wireless Data for the Enterprise*

Reference
MULLER • *Desktop Encyclopedia of Telecommunications*, 3/e
BOTTO • *Encyclopedia of Wireless Telecommunications*
CLAYTON • *McGraw-Hill Illustrated Telecom Dictionary*, 3/e
RADCOM • *Telecom Protocol Finder*
PECAR • *Telecommunications Factbook*, 2/e
RUSSELL • *Telecommunications Pocket Reference*
KOBB • *Wireless Spectrum Finder*
SMITH • *Wireless Telecom FAQs*

Security
HERSHEY • *Cryptography Demystified*
NICHOLS • *Wireless Security*

Telecom Engineering
SMITH/GERVELIS • *Cellular System Design and Optimization*
ROHDE/WHITAKER • *Communications Receivers*, 3/e
SAYRE • *Complete Wireless Design*
OSA • *Fiber Optics Handbook*
LEE • *Mobile Cellular Telecommunications*, 2/e
BATES • *Optimizing Voice in ATM / IP Mobile Networks*
RODDY • *Satellite Communications*, 3/e
SIMON • *Spread Spectrum Communications Handbook*
SNYDER • *Wireless Telecommunications Networking with ANSI-41*, 2/e

BICSI
Networking Design Basics for Cabling Professionals
Networking Technologies for Cabling Professionals
Residential Network Cabling
Telecommunications Cabling Installation

Developing MMS Applications

Developing MMS Applications

Multimedia Messaging Services for Wireless Networks

Scott B. Guthery

Mary J. Cronin

McGraw-Hill

New York Chicago San Francisco Lisbon
London Madrid Mexico City Milan New Delhi
San Juan Seoul Singapore Sydney Toronto

The McGraw-Hill Companies

Cataloging-in-Publication Data is on file with the Library of Congress

1 2 3 4 5 6 7 8 9 0 DOC/DOC 0 9 8 7 6 5 4 3

ISBN 0-07-141178-X

The sponsoring editor for this book was Stephen S. Chapman and the production supervisor was Pamela A. Pelton. It was set in New Century Schoolbook by Patricia Wallenburg.

Printed and bound by RR Donnelley

McGraw-Hill books are available at special quantity discounts to use as premiums and sales promotions, or for use in corporate training programs. For more information, please write to the Director of Special Sales, McGraw-Hill Professional, Two Penn Plaza, New York, NY 10121-2298. Or contact your local bookstore.

This book is printed on recycled, acid-free paper containing a minimum of 50% recycled, de-inked fiber.

Dedicated with thanks to everyone at Mobile-Mind

CONTENTS

Foreword xv

Acknowledgments xix

Chapter 1 The Move to Multimedia Messaging 1

In Search of Applications 2
Mobile Application Platforms 3
Inside MMS 6
MMS in Action 6
Comparison of MMS and SMS Features and Limitations 12
Comparison of MMS and WAP 13
Getting Started with MMS Development 15
Essential Resources for MMS Development 16
Summary 17

Chapter 2 Understanding the MMS Architecture 19

MMSC Connection to the Mobile Handset 21
MMSC Connections to Other Computers 23
MMS Media Elements 25
MMS Message Assembly 26
MMS Message Passing 28
Communication Bearers and Protocols 30

The MMS Model of Operation 32
Summary 33

Chapter 3 Structure of an MMS Message—The MM1 Interface 35

Request and Response Envelopes on the MM1 Interface 37
The Basic MMS Model of Operation 40
The MM1_submit Envelope 41
X-Mms-Message-Type 43
X-Mms-Transaction-Id 43
X-Mms-3GPP-MMS-Version 43
X-Mms-Recipient-Address 43
X-Mms-Sender-Address 44
X-Mms-Message-Class 44
X-Mms-Date-And-Time 44
X-Mms-Time-of-Expiry 44
X-Mms-Earliest-Delivery-Time 45
X-Mms-Delivery-Report 45
X-Mms-Reply-Charging 45
X-Mms-Reply-Deadline 45
X-Mms-Reply-Charging-Size 45
X-Mms-Priority 45
X-Mms-Sender-Visibility 46
X-Mms-Read-Reply 46
X-Mms-Subject 46
X-Mms-Reply-Charging-Id 46
X-Mms-Content-Type 46
MIME Encapsulation 47
The SMIL Program 49
SMIL 2 MIME 51
MM1 Submit of the "Hello, World" MMS Message 54
The Response of the MMSC 54
X-Mms-Request-Status 55

X-Mms-Request-Status-Text 55
X-Mms-Message-Id 56
Notification of the Arrival of the "Hello, World" MMS Message 56
Delivery of the "Hello, World" MMS Message 60
Other MM1 Envelopes 63
A Cautionary Word about Transcoding and Content Adaptation 64
Summary 68

Chapter 4 MMS Development Tools and Resources 71

MMS Conformance SMIL 76
MMS Construction 79
Compose 79
Transform 81
Encode 83
SMIL Players and MMS Handset Simulators 84
Application Developer Programs 85
Other Useful Software Tools 87
Summary 88

Chapter 5 Media Formats 89

Determining What a Handset Handles 96
Text 104
Audio 108
Graphics 114
Miscellaneous 118
Summary 119

Chapter 6 Testing Your MMS Application 121

MMS Portals and Web-Based MMS Composers 123
MMS Application Developers' Programs 124

Direct Loading of an MMS into the Handset 124
Sending an MMS from Your Own Server 125
ASCII versus Binary Representation of MMS Messages 126
The Binary Representation of the "Hello, World" MMS Notification 127
An MMS Notification as a WML Page 130
The Binary Representation of the Hello, World" MMS Message 131
An MMS Notification as an SMS Message 136
An MMS Notification as a WAP Push 143
Summary 145

Chapter 7 Using MMS for Mobile Delivery of Content—The MM3 Interface 147

E-Mail and MMS 149
RTP and MMS 151
SMIL Templates and Automatic MMS Production 153
Boston Bruins Example 156
RTP Example Summary 157
Databases and MMS 157
MMS Message Store 158
Summary 160

Chapter 8 The Value-Added Service Provider Interface—The MM7 Interface 163

Submit on the MM7 Interface 165
MM7 Messages 170
Copyright Protection and Digital Rights Management Issues 173
Security 174
Summary 174

Chapter 9 The MMS Business Case 177

The MMS Value Chain 179
Strategic View of Value Chain Participants 180
Pricing Fundamentals 182
Application Developer's Perspective 184
Getting Applications to Market 185
Matching Pricing and End User Value 186
Real-World MMS Distribution Paths 189
Bringing Developers on Board in Singapore 189
Adding MMS to the BREW 191
Summary 192

Chapter 10 Next Steps in Multimedia Messaging 195

Issues to Watch 196
Content Protection and Digital Rights Management 196
Interoperability of Hardware and MMS Standards Conformance 198
Roaming Between Operators and Between Network Technologies 200
Value-Based Pricing for MMS Applications 201
The Role of the MMS VASP 202
Interaction with Technical Substitutes: Polyphonic Ringtones, EMS with SVG, IMS 203
Expanding to New Markets 204
Sponsored MMS and MMS Advertising 204
Enterprise MMS 205
Conclusion 206

Appendix A MMS Standards and Specifications 207

Network Standards and Specifications 208
Media Standards and Specifications: MMS-Specific Usage 210

Media Standards and Specifications: Individual Media Formats 210

Appendix B A Complete MMS Message on the MM1 Interface 213

Appendix C UAProf Schema for MMS Handsets 225

Index 261

FOREWORD

"One day, every town will have a telephone."

This is reputed to be a quotation from around 1876, the time of Alexander Graham Bell.

With the advent of mobile telephony during the last decade of the twentieth century an appropriate quotation today would be that "One day, every person will have a mobile telephone."

The way we conduct our lives has been changed dramatically since those humble beginnings in 1876.

Historically, telephones have been associated with a place but mobile telephony allowed telephones to be associated with a person. The difference is that a person is now potentially and virtually instantaneously capable of being contacted no matter where they are in the world. The impact has been far reaching for business and pleasure and has even had some undesirable cultural impacts, as phones become a nuisance factor in public places. But, mobile telephones are here to stay.

Mobile telephones were never intended to compete with the "local loop" but today the situation is very different. We see people preferring to use a mobile telephone rather than their conventional 'landlocked' telephone. It is the convenience factor that is important to them and conventional telephones have struggled to compete with the sophistication of mobile phones despite the cost. The future for the landline into the office and home seems to be destined for high-speed digital communications.

Mobile phones also developed and introduced the public to features such as messaging that are not commonplace on conventional telephones. The beginnings of mobile telephone messaging began with the Short Message Service, which was first conceived in the late

1980s and by the early 1990s was operational across most GSM networks. Who would have thought that a service that took at least 5 seconds to deliver a 160 alpha—numeric character message entered from a mobile phone 0 to 9 digit keypad (using the thumb) would have resulted in the next largest revenue earner to speech? The attraction of SMS to youngsters and teenagers was totally unforeseen—it was fun. A new shorthand language had developed, e.g., CUL8A (see you later). SMS was certainly not cheap, 10 pence UK being the typical cost for a single message irrespective of the number of characters sent.

By the late 1990s the revenue from SMS was showing signs of reaching a plateau. Users, particularly the youngsters, were beginning to look for a new messaging experience. In response to this, Nokia introduced "picture messaging" which was a great success provided both parties had a Nokia phone. Sadly this was not always the case. This fact seized the attention of the GSM standards committee who then developed the specifications for enhanced messaging that allowed polyphonic sounds and vector graphics to be sent within a single or within a very few short messages. It was now possible to send simple animations. Enhanced messaging products were slow to appear, primarily because manufacturers and network operators were focussed on a bigger agenda item—MMS.

The expectations set for MMS were high and remain so to this day. MMS brings the promise of still and moving pictures, complex sounds and moreover the Internet in your pocket. What has emerged is an obvious need to add the stimulus of visual communications to messaging. That today seems to be the overriding interest and is developing rapidly.

The challenge facing MMS is the expectation of its alignment with the Internet with regard to performance (e.g., speed), commercial considerations (e.g., tariffs) and operational considerations (addressing—telephone numbers vs name@domain). This will require network operators to manage people's expectations because in contrast to the Internet, mobile call costs are high and data rates are slower. However, the advantage is mobility. This advantage may face further challenges with the emergence wireless LANs that have the benefit of low cost and high speed as well as a degree of mobility that might satisfy some applications. Nevertheless, the mobile phone is bound to remain the prime mechanism for mobile communications for a long time to come.

Mobile telephony will no doubt go down in history as one of the greatest achievements of the twentieth century. Mobile multi media messaging, in whatever form that materialises, will surely go down as one of the greatest achievements of the first decade of the twenty-first century.

This book should prove to be an extremely valuable tool for those having an interest in promoting MMS business opportunities and for application developers. It is an easily read interpretative supplement to the formal and complex messaging specifications. Typical "use case" examples are described to which the reader will find it easy to relate. The closing pages of the book address the emotional yet serious issues of interoperability, digital rights management, and roaming from a realistic and pragmatic perspective.

Ian Harris, CEng, FIEE

(Ian Harris is a Chartered Engineer and Fellow of the Institute of Electrical Engineers. He is currently the Chairman of 3GPP TSG T WG2 [Mobile Terminal Services and Capabilities] that includes the MMS messaging capabilities of 3GPP terminals.)

ACKNOWLEDGMENTS

The Multimedia Messaging Service (MMS) provides a powerful and exciting new opportunity for mobile application creation and delivery. By creating a standard format for combining images, voice, and text formats and by harnessing a wireless network infrastructure for delivering multimedia applications to a variety of mobile handsets, MMS opens the door to rich media content for mobile developers. However, the relatively recent emergence of MMS standards, handsets, and distribution systems combined with the lack of consistent documentation about designing and distributing MMS applications, makes MMS development a challenging frontier. We hope that *Developing MMS Applications* will provide some helpful advice and guideposts about the technical foundations and the business case for MMS. Given how quickly MMS is evolving, we recognize that some of the information provided in this book will need updating. To check on the latest developments, please visit our MMS website at http://www.mobile-mind.com/htm/mmsupdate.htm.

As always, we have relied on the volunteer and largely unsung work of the international telecommunications standards groups, particularly 3GPP, 3GPP2, and OMA, as the definitive source for the technical specifications that underpin the examples in this book. We would like to thank 3GPP and ETSI for permission to cite examples and reproduce tables. In addition, we are especially grateful for the careful critical reading of the entire manuscript and numerous suggestions for improvement from Vjekosla Nesek and Bryce Norwood.

Closer to home, many thanks to Matt Kowalewski of Mobile-Mind for his invaluable help with graphics, tables, and close reading of the content. Without the contributions and support of all the great people at Mobile-Mind, this book would have taken much longer to produce. We thank you.

Developing MMS Applications

CHAPTER 1

The Move to Multimedia Messaging

In Search of Applications

More than a billion people worldwide own a mobile phone, a number that dwarfs the number of personal computer (PC) owners. Market leader Nokia projects sales of between 50 to 100 million color screen, multimedia capable phones in 2003, with mobile phone shipments from all vendors expected to top 400 million.[1] This compares to projections that consumer and business users combined will buy fewer than 150 million new PCs during 2003.

In Italy and Sweden, there are four mobile subscribers for every PC owner. In Japan, more than 75 percent of the population currently owns a mobile phone, and wireless Internet connections are far more popular than online PCs. In fact, the United States is one of the few countries where more citizens own a computer than a mobile phone, and even in the United States, the adoption of new mobile devices is growing much more quickly than the sale of new PCs.[2]

At the same time that mobile phones have raced ahead of PCs in terms of global adoption, the typical cell phone has developed into a full-fledged computing platform capable of handling data and running multiple applications. Today's advanced mobile phones support operating systems that are compatible with Java (J2ME) and Windows. The typical mobile device purchased in 2003 has more processing power than the first generation of PCs that revolutionized end user computing in the 1980s. A comparable revolution in mobile subscriber expectations and behaviors seems inevitable.

A look at the number of applications available to PC users compared to applications written for the mobile phone, however, shows that PCs still dominate this critical area. PC user applications number in the hundreds of thousands, and the sale of PC applications is a multibillion dollar global industry. In contrast, only a few thousand mobile phone applications are available through commercial distribution, and these have been concentrated in a handful of countries, primarily Asia and Northern Europe.

---

[1] Nokia Press Release, "Nokia expects solid growth in the mobile handset industry in 2003," December 3, 2002. http://press.nokia.com/PR/200212/883734_5.html

[2] "Nokia v Microsoft, the fight for digital dominance," *Economist Survey*, November 21, 2002.

Clearly, a mobile applications gap exists. Despite rapid advances in mobile device capabilities, the mobile world lags behind the PC industry in application development and sales. This gap is causing problems for wireless carriers who are counting on the revenue from value-added applications to offset the declining prices for wireless voice communication, and the considerable costs of upgrading to higher speed network infrastructures worldwide.

Multimedia messaging service (MMS) is a leading contender to fill this mobile applications gap, with products that millions of wireless customers will find entertaining, exciting, worth downloading to their phones, and perhaps most important, worth the expense. Some analysts are already predicting that within three years, MMS applications and revenues will overtake the popularity of short messaging service (SMS) messaging, and will eventually become the dominant mode for delivering mobile applications to consumers. Other observers, while agreeing that MMS offers great potential, are skeptical about its mass adoption worldwide, citing the need for special handsets and unresolved issues, such as interoperability and roaming.

This book is designed to bring application developers up to speed on the technical requirements for MMS application development and delivery. In addition to mastering the tools for building MMS applications, the savvy developer needs to understand how those applications will enter the distribution chain. So, *MMS Wireless Application Development* will provide a step-by-step technical description of the path that an MMS application takes, from the developer's workstation to the mobile subscriber's phone, as well as a business overview of the MMS value chain and analysis of how each business participant views (and values) its part of the action.

Mobile Application Platforms

In the search for mobile applications, MMS is only one of the areas where carriers and developers are placing their bets. So it is important to understand how MMS application development compares to the other alternatives. To reach mobile customers today, there are three viable platforms on which to build applications: an Internet server, the handset, and the subscriber identity module (SIM). Table 1.1 lists some of the differences among these three platforms, from an application developer's point of view.

TABLE 1.1
Mobile Application Platforms

	Internet Server	Handset	SIM
Interface technologies	SMS, EMS, MMS, WAP, IMS, HTTP	MExE, BREW, J2ME, Symbian, ExEn, Mophun	SIM Toolkit
Barrier to entry	Low	Medium	High
Access to communication channels	Low	Medium	High
Interaction with network operators	Low	Medium	High
Memory resources	Megabytes	10 to 200 Kilobytes	1 to 20 Kilobytes
Processor resources	100s of MIPS	10s of MIPS	500 KIPS
Typical application	Messaging	Games	Payment
Number of applications and application developers	100,000s	1,000s	100s
Dominant players	Infrastructure providers (e.g., Logica, CMG, etc.)	Handset providers (e.g., Nokia, Ericsson, Panasonic, etc.)	Network operators (e.g., Vodafone, T-Mobile, etc.)

From a developer's perspective, aspects of any given application could be slotted into each of these columns. Developers may want their application to be in regular communication with the end user for updates, new information, etc., so there is a messaging aspect. If the top priority is very predictable performance and quick turns when end users are interacting with the application, the best platform may be the handset. And if there is a high security requirement for the application, it may be desirable to leverage the physical security token owned by the carrier, the SIM for example, to uniquely identify the user engaging in a mobile financial transaction, or even to get paid.

For a number of reasons that are much more historical and business-driven than technical, these three platforms are not well integrated today. The unfortunate result is that building a distributed mobile application that sits on all three platforms and harnesses and synchronizes the features of each is at best difficult and at worst simply impossible. Developers need to select a target, such as MMS on the handset, and work with its strengths and shortcomings to bring their applications to the marketplace.

The other business reality is that application providers are not dominant players on any of the three platforms. The platforms are designed, standardized, and evolved primarily with proprietary interests in mind, not for the ease of use or the benefit of the application and content providers who create end user value from the platforms.

This can be a source of frustration for application developers coming from other computing venues such as Web design, database systems, or embedded software, where the information needed to build applications is widely and freely available, and platform providers cater to application developers. The historic attitude in telecommunications has been that the platform providers can easily supply the most important applications themselves, rather than encouraging and supporting third-party application developers and content owners to share in revenues.

MMS is forcing a change in that attitude. Content is intrinsically expensive to create, labor intensive to maintain, and driven by the whims and fads of the marketplace, not the dictates of telecommunication engineers and standards bodies. By and large, the current telecommunications leaders do not have any core expertise in content creation and management, and a number of traditional operators have a history of failed attempts at merging content and communications services on the Internet. However, content and the applications that leverage content are clear requirements for success in the MMS world. This content and application requirement is driving a new appreciation of application developers and their importance in the adoption of MMS by wireless subscribers.

This is a monumental change that has only begun to happen, and many of the old attitudes and ways of doing mobile and cell phone business are still firmly in place. Developers may still run into barriers in obtaining the information needed to build MMS applications. Specifications for MMS handsets and the support for specific messaging formats may change unexpectedly for no reason. Application interfaces that worked well yesterday may be unavailable on the latest MMS-capable handsets. Just as the carriers and infrastructure providers have to build bridges to the application developers and content owners, developers must be ready to work closely with the leading carriers, device makers, and infrastructure providers to assure that their applications can be distributed to the widest possible market. Developers who know MMS from the inside out will be positioned to make the infrastructure work for them—and for their customers.

Inside MMS

MMS supports the delivery and display of multimedia content on MMS-capable mobile devices. The content of an MMS message can include color graphics, animation, sound, text, and even video, as well as special-purpose formats such as vCards and vCalendars. These capabilities open the door to varied applications, from delivering personalized mobile greeting cards to a sports highlights subscription service or a selection of animated cartoons for consumers. Using an MMS phone with an integrated digital camera, a mobile subscriber can snap a picture, add a text message, and send it directly to a friend who also has an MMS phone. While MMS messages can be created on the phone itself and exchanged among subscribers, more sophisticated applications require the efforts of developers and content providers to take full advantage of the still limited phone display. One typical MMS applications is illustrated on a phone screen in Figure 1.1.

MMS in Action

Take a closer look at how the personal photo image, as well as animated comic strip and subscription sports service, illustrated in Figure 1.1 turns into an MMS message and pops up on the screen of a

Figure 1.1
MMS personal photo.

mobile phone. We will first describe the mobile subscriber's point of view in using various MMS applications, then highlight some questions about the relationships and network links that are likely to surface in implementing each scenario.

Mobile photo. Peer-to-peer messaging has been the driving force behind the phenomenal growth of SMS around the world, with billions of SMS text messages exchanged each month among mobile subscribers. Peer-to-peer exchange of mobile phone-generated photos is already a popular activity for subscribers in Japan and Korea. Japan, however, has the advantage of a rapid adoption of photo-capable phones and a critical mass of subscribers using the same network. In the United States and other countries, it is expected that a slower adoption rate of MMS phones and lack of interoperability among multiple wireless networks will slow down peer-to-peer MMS messaging unless alternatives, like online access to MMS messages, are available. Scenario 1 reflects some of the issues facing peer-to-peer deployment of MMS photo exchange in the United States.

Scenario 1

Amy is eager to try out the camera attachment for her Sony Ericsson T68i MMS phone. She snaps pictures of everything in sight, including her college roommates. However, taking the pictures is only step one. The main reason for attaching a camera to a phone is being able to send a photo instantly to someone who is interested in seeing the picture and the text message that Amy can send along with it. Should she send a picture to her roommates' parents? How about to her best friend in high school who is now attending college in California? In selecting recipients for her MMS photo message, Amy needs to consider the following questions:

- Does the recipient also have an MMS-capable phone?
- If yes,
  - Is that phone connected to the same wireless network that provides her service?
  - If the recipient has a different wireless carrier, is there an MMS roaming agreement in place between the two carriers?
  - Will Amy pay for the full cost of delivery of the message to the recipient's phone, or will a charge show up on the recipient's phone bill?
  - Does she want to request a receipt of delivery so that she will know when the recipient gets a notification that there is a message ready for downloading?
    - Does she also want to know when the recipient actually opens the message and views the photo?
- If no,
  - Is there a way for Amy to send the photo to an Internet-accessible server so that the recipient can log in via the Web to view the message and photo?
    - Will this service generate an SMS or e-mail notification to the recipient that includes instructions for viewing the photo and message via the Web?
    - Can she send this notification to multiple recipients?
    - How long will her photo and message stay on this server?
    - Is there an extra charge for this notification?

As described in later chapters, there is no universally applicable answer to Amy's questions. Wireless carriers and MMS infrastructure providers are still experimenting with business and service models, as they move from launching MMS services to establishing partnerships and revenue targets. Even the most basic scenario of an MMS subscriber taking a personal photo with his or her phone and wanting to send it to a friend requires the pre-existence of a number of business and technical relationships.

When commercial content—such as an animated cartoon strip or a subscription to sports highlights—is part of the MMS service, the scenarios are different, but equally complex.

Original animated comics. In addition to digital images with text captions, MMS applications can handle sound and animated special effects. With the appropriate tools and graphic talents, an MMS developer can create new types of original entertainment for mobile customers. One entertainment format that adapts well to the mobile environment is the animated comic strip or short cartoon. MMS animations today range from the purely silly (and occasionally even funny) to surprisingly realistic adult entertainment offerings. These may be available on MMS-related Web sites free for download and exchange, or may be restricted to the paying subscribers' phones. Either way, the introduction of third-party created graphic content adds another layer to the MMS development and distribution value chain.

Scenario 2

Erik is bored with SMS jokes and horoscopes, but he is an avid fan of Ductman, the irreverent hero of his daily MMS animated comic strip. He happily pays a monthly subscription fee to start his work days with a 60-second dose of Ductman's struggles, stratagems, and triumphs that somehow echo Erik's own contrarian view of the world. In fact, Erik saves the weekday episodes on his phone for a final Friday night viewing. It doesn't bother him that Ductman comics are strictly limited to his MMS phone. In fact, this exclusive distribution channel (exclusive in comparison to the Web, at least) adds to his sense of being part of a small "insider" group who can relate to Ductman's antics. Erik would like to message the Ductman creators with suggestions and comments, and occasionally he thinks it would be interesting to have a way to compare notes with other Ductman fans. However, he mainly wants Ductman to keep his edge and avoid all the baggage of mainstream media content.

Distributing original content via MMS requires a front end for subscribers to subscribe and pay the required fee, as well as decisions about the level of content protection and subscriber services. Among the questions that content providers and application developers must resolve are:

- Will distribution and viewing of MMS original content be limited to the subscribers' phones or also downloadable to personal Web pages and PCs?
- Should the content be developed with a built-in expiration date when it will no longer operate even on the phone, or should it be permanently accessible to the subscriber?
- What techniques are best suited to provide the desired level of content protection with the least impact on MMS message size, download speed, and runtime performance?
- Is the marketing and distribution of niche MMS content such as Ductman comics best done through independent MMS portals, through wireless carriers, or by the content creators?
- Who will pay what fees in the process of moving the Ductman comic MMS from the content creator to the end subscriber's handset?

Many of the content protection and digital rights management questions also apply to MMS content services that are based on existing online or broadcast media. However, there are important differences in the back-end content conversion and MMS infrastructure requirements, as well as in business models and opportunities for value-added services in settings like the MMS-based distribution of professional sports highlights, as described in Scenario 3.

Professional sports highlights. With a pre-existing mass market of fans, intrinsic graphic appeal, and billion-dollar broadcast agreements already in place, professional sports highlights are a natural source of MMS content. There is, of course, a dramatic difference between viewing game-winning shots on a wide screen television and a tiny mobile phone. Broadcasters and sponsors have decades of experience in shaping and meeting TV audience expectations for sports information, and there will likely be a parallel period of experimentation and learning before the appropriate media feeds for mobile viewing are established. Rather than expecting end users to pay for MMS sports updates, sponsorships may play a key role in driving adoption.

Scenario 3

Matt's MMS phone plays a familiar commercial jingle as the ring tone that announces the arrival of the day's professional sports highlights. When Matt accepts the incoming MMS message stream, the first icon on his phone screen is the logo for the soft drink maker associated with the jingle. Normally, Matt would be annoyed to receive advertising on his phone—but these messages are an exception. He knows that without the sponsorship, he would be paying a hefty monthly bill for the daily MMS messages that show winning plays and contested calls for his favorite teams. By dealing with the commercial introduction, he gets the service for free. That's a deal he is happy to make.

The transition from broadcast highlights to sponsored mobile distribution raises a number of issues that need to be resolved before the potential of MMS sports services can be fully realized. Among these are:

- Security of digital rights management (DRM) solutions that will prevent copying and secondary distribution of the content.
- Variable value and pricing points for sponsorships of real-time highlights for high-stake sports events (play-off series, world cup, and championship matches), compared to sponsored clips of hometown teams in action.
- Infrastructure required to capture real-time media feeds and create MMS messages "on the fly" for time-sensitive content.
- Ownership of content and editorial commentary (sportscasters' analyses, players' comments, etc.) that might accompany the MMS message.
- Convergence with existing online sports services that distribute game highlights and statistics via the Web to create all-media sports portals, or separate agreements between teams and carriers for MMS distribution.
- A fan's desire and willingness to pay a premium for value-added mobile features and services that are not covered by sponsors.

These three scenarios illustrate the varied opportunities for MMS services, as well as the multiple questions and issues posed by each new service. MMS is a new technical and business arena, as different from SMS text applications as it is from the development of online content services. Yet, MMS has similarities with both SMS and the Web. The next section will compare MMS to SMS and wireless application protocol (WAP) application development.

Comparison of MMS and SMS Features and Limitations

The enormous popularity of SMS came as a surprise to wireless carriers. While billions of SMS messages are exchanged each month in the form of simple text messaging among subscribers, there are also thousands of SMS developers who have stretched this simple format into a platform for games, news and entertainment services, and countless other applications worldwide. Given the much richer graphic features of MMS, carriers expect that these same developers will be even more creative in building thousands of popular multimedia applications. Wireless analysts predict that by 2007, SMS traffic will decline significantly, as MMS becomes the preferred medium of exchange for both peer-to-peer messages and commercial applications. To evaluate this prediction, it is helpful to review the key features and limitations of both SMS and MMS. Table 1.2 provides an overview of these.

TABLE 1.2

Comparison of SMS and MMS

	SMS	MMS	Comments
Delivery basis	Store and forward	Store and forward	Via SMS centers or MMS centers connected directly to the carrier network
Formats	Text	Graphics Audio Text captions Vcards, calendar, etc.	

(continued on next page)

TABLE 1.2

Comparison of SMS and MMS (continued)

	SMS	MMS	Comments
Size limits	160 characters of text (most carriers allow chaining of two or more messages)	30K or more of data (specific limits defined by carrier)	
Interactivity (two way messaging options)	Yes	Yes	
Device requirements	Minimal—any text-capable handset	Specialized MMS-compatible handsets	
Availability of devices	Mass production worldwide	Selected carriers and manufacturers	
Cost to subscriber	Free (bundled with calling plan) or low cost (pennies) per message	Significant cost (up to $1) per MMS message sent, different rates for incoming messages	Many carriers are providing free MMS messaging during launch period for the service

Comparison of MMS and WAP

One concern about the success of MMS applications is that a number of the same contributing factors that delayed or derailed earlier contenders in the mobile applications arena, such as WAP, may also hinder MMS adoption. These factors include:

- Inability of older mobile devices to access the service, requiring subscribers to upgrade to expensive new handsets just to evaluate the applications
- Lack of interoperability between mobile phones and between different networks requiring device-by-device testing to ensure comparable performance of applications in different environments
- “Walled garden” strategy of wireless carriers to limit access to proprietary or closely controlled application offerings by a few selected vendors

- Lack of business models for sharing revenue with application developers
- Resultant lack of a broad variety of applications to drive mass market adoption

However, MMS has some technical and business advantages over both SMS and WAP that make the predictions of billion-dollar MMS revenue generation more likely to come to pass. It is useful to consider these advantages from the viewpoint of different participants in the MMS value chain, as summarized in Table 1.3.

TABLE 1.3 Advantages of MMS

	Advantages of MMS
Carrier	■ New revenue source and opportunity for differentiation from commodity competition on cost of calling plans ■ Justification for investment in higher speed network infrastructure ■ Opportunity to develop value added services based on content and development partnerships
Device maker	■ Driver for adoption of new, high-end devices and replacement of older devices ■ Elevates role of device as platform for MMS services and possible solution to content protection and application management
Content provider	■ New outlet for existing content, with protection against the free copy and forward mindset of the Internet ■ Opportunity for subscription-based services, global distribution of content, and revenue sharing with carriers
Application developer	■ Standards-based development environment ■ Open and emerging field for new partnerships and business models ■ Access to mass markets worldwide
Subscriber	■ More value from the mobile device ■ Instant availability of entertainment and other services

The business case for MMS rests with these value chain participants. Their relationships and strategies for getting applications to market are described in more detail in Chapter 9.

Getting Started with MMS Development

Two primary technical aspects of MMS application development require in-depth discussion. The first technology to master is the construction of the multimedia message itself. The second essential MMS component that developers must understand is the available options for sending an MMS message into the network and getting it to the user or customer. These two aspects of MMS development will form the core of this book.

Wisely, the mobile network industry has not tried to create all of its own media formats, so the technologies you will be working with in constructing an MMS message will be old friends such as WAV, JPEG, MIDI, GIF, and ASCII. If you have been active in multimedia, you may already be familiar with SMIL, the layout language, and MIME, the packaging protocol that MMS uses to tie together and coordinate the components of a message. If you developed applications for the Web, you know about the HTTP and SOAP protocols.

To be sure, you will have to learn to deal with some constraints and limitations on the use of these tools that are a consequence of the fact that your message is being sent through a relatively low-bandwidth wireless communication channel to a playback device of only modest multimedia playback capability. Working with the small displays and limited color palettes of mobile phones is a challenge for every graphic artist, as is the audio subsystem on the mobile phone, which has been hyperoptimized for human speech.

You will also probably become more conscious of the size of your creations than you have been to date, and will find yourself making sure that each media element you add to your message is worth what you are paying to get it to your customer. If the overall message can only be 30K, then you want to make sure you are getting the biggest bang per byte that you can.

Once you have built your message, you are ready to send it to your user. We will explore some of the message encodings that are unique to the mobile world so you have at least a passing knowledge of them. A good model to have in mind is that launching an MMS message is not much different than launching a Web page. MMS includes a group of hypertext transfer protocol (HTTP) headers that tell the network operator about the message and what to do with it (like who to send it to and when it has to arrive), together with the message itself in the body. For the most part, it walks and talks like an e-mail, except the address is a phone number.

Essential Resources for MMS Development

The two basic resources you will need to get started with multimedia messaging application development are a mobile handset that supports GPRS or MMS, and an account on a mobile network that supports MMS transmissions.

The Sony Ericsson T68i handset is to MMS application development what the Nokia 5190 was to SMS application development: the developer's warhorse. However, there are many MMS capable handsets available, and new options are hitting the market daily. Make sure that the development handsets you buy can play back the media formats with which you want to work. More detail about handsets is in Chapter 5.

When choosing a network operator, try to find one that covers your development environment with a good, strong signal, and one that doesn't believe in "walled gardens." You will want to interact with developers connected to other operators, and will want to beta test your MMS application with friends who probably don't want to change their mobile telephone number just to help you out.

Seeing the MMS messages you develop on a real handset is absolutely critical. Don't stop with handset simulators and multimedia players that you run on your workstation. The hardest part of MMS design and implementation is dealing with the stark reality of the handset. Simulators and players don't force you to step up to the hard problem. Only a real handset does that.

Currently, no integrated software development environment for MMS messages like Visual Studio or Cold Fusion exists. You will find yourself working with a large number of individual tools. A good image editor such as Paint Shop Pro, and a good audio editor such as GoldWave are "must haves." You have to be able to "sweat the bits" when you're building an MMS application so you have to get tools that let you work with the bits when needed.

If you are building a commercial MMS application, at some point you will want to move beyond sending your MMS messages through a mobile phone hanging off the side of your workstation. This means getting a direct connection to the mobile network either directly from a network operator or indirectly through an MMS broker or MMS value-added service provider. The sooner you step up to finding this connection the better, because your development will go faster and you will be working out the technical details of distributing your application at the same time you are developing the application itself. Plus, you may find that your high-volume connection provider is also a potential business partner for distributing your application.

The starting point for building MMS applications is the architecture and the interfaces of the multimedia messaging service. These are the topics of Chapter 2. Before delving into details of MMS architecture, take a brief look at the topics this book will cover.

Summary

After this close-up look at the advantages of MMS for application development and a comparison of SMS and MMS in Chapter 1, Chapter 2 provides essential information about the architecture and application developer interfaces that underpin MMS development and message distribution.

Chapter 3 guides the developer through the process of building a simple MMS application from the ground up, then preparing it by hand for the round trip through a multimedia messaging service center (MMSC), over the wireless network, and back to the mobile phone. The details of this process provide a context for Chapter 4's discussion of using software tools and infrastructure resources to create complex MMS messages for commercial distribution.

Chapter 5 delves into the MMS media formats, including image formats such as GIF and JPEG, audio and video formats, and an overview of currently available MMS handsets and development features. Testing applications is an essential development step, and Chapter 6 reviews testing options, from MMS portals to building your own testing environment. Chapter 7 examines the options for using MMS for delivering existing media content, including integrating MMS services with databases and real-time data feeds, while Chapter 8 describes the technical interfaces and application management challenges of the MMS value added service provider (VASP).

Chapter 9 provides an in-depth look at the business case for MMS application development, with an analysis of the MMS value chain, pricing and billing issues, business partnerships, and the role of application developers. Chapter 10 concludes with a preview of the next steps in MMS evolution, including the drivers for and barriers to adoption of advanced MMS applications. This final chapter summarizes the essential technical and business factors for the successful development and deployment of MMS, including digital rights management (DRM) for MMS, roaming agreements among carriers, and the challenges of interoperability.

CHAPTER 2

Understanding the MMS Architecture

The basic elements of any messaging architecture are a sender, a router, and a receiver. In the postal system, the post office serves as the router of letters between sender and recipient. In the short messaging service (SMS), the short messaging service center (SMSC) is the router between the SMS sender and the SMS recipient.

In many messaging systems, the router is a store-and-forward function that may hold the message awaiting some event before moving it on. In the postal system, the event could be the arrival of the postman who picks up the bag of letters for the residents on the street where the recipient lives. In the SMS system, the event is the appearance of the recipient's phone on the network.

Two aspects of MMS make it a unique messaging service in the galaxy of telecommunications messaging services. First, the messages are large, as compared to the messages carried in other services. Second, it is expected that many of the messages will be generated by network services and servers and batch distributed, rather than prepared and sent to and from individuals.

As described later, the size of the message necessitates an extra step in delivery wherein the recipients are notified of the arrival and availability of an MMS message and asked if they want it delivered. This is unlike SMS messages, which are delivered automatically without prior consent. The size of MMS messages also dictates new policies and protocols within the network, since storing millions of MMS messages awaiting their forwarding event entails a nontrivial amount of computer resources.

The second feature of MMS messages, the fact that they are coming from network servers and not just individuals, introduces an asymmetry in the MMS system: sources of messages are different from sinks. Furthermore, there are many different kinds of sources and really only one kind of sink—namely the subscriber carrying an MMS handset.

Figure 2.1 is the MMS reference architecture taken directly from the 3GPP standard that defines how MMS works, 3GPP TS 23.140. This architectural diagram exhibits the impact of the two unique aspects of the MMS messages described previously.

In the center of this diagram is the MMS Relay/Server. This architectural component is also called the MMS Proxy-Relay in some specifications. The MMS Relay/Server plays roughly the same role in

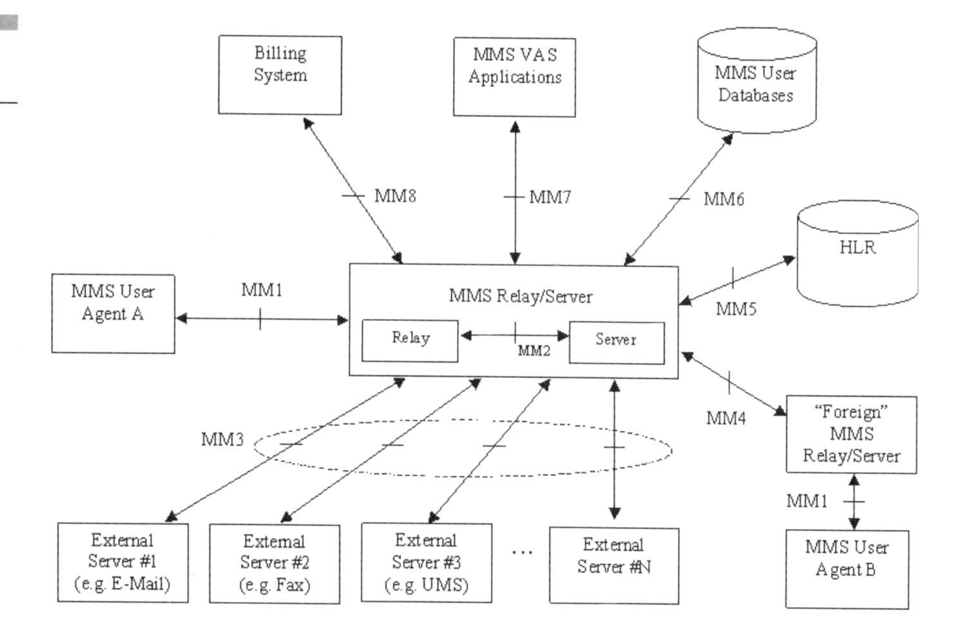

Figure 2.1 MMS architecture.

the MSM as the SMSC does in SMS. The MMS Relay/Server is often referred to as the multimedia messaging service center (MMSC). We adopt this convention in this book.

All multimedia messages pass through at least one MMSC. If the sender and the recipient are both connected to the same MMSC, then that MMSC only is involved in getting the message from where it originated to its destination. If the sender and the recipient are connected to different MMSCs, then there will obviously be at least two MMSCs involved; one through which the sender submits his MMS message to the network, and a second one from which the recipient receives the MMS message. There may be other intermediary or relay MMSCs involved in getting the message from the sender's MMSC to the recipient's MMSC.

MMSC Connection to the Mobile Handset

Look at the left-hand side of Figure 2.1 and, in particular, the box labeled "MMS User Agent A." User Agent is standards talk for a mobile handset. This is where a mobile handset, like a Sony Ericsson T68i mobile telephone, connects to the MMS network. The expression "User Agent" is used to provide a little generality to the specification because all that is really needed here is a device that can either create an MMS message or play one back. In the future, this could be a personal digital assistant (PDA), a desktop computer, a television set, or a wholly new multimedia playback device that hasn't been invented yet.

The diagram shows that a mobile handset, like the Sony Ericsson T68i, connects to the MMSC using an interface called MM1. Using MM1, the handset receives multimedia messages and sends multimedia messages to the MMSC.

If we think of a multimedia message as a big blob of bits—which it is after all, and we will tear apart and describe it in detail in Chapter 3—then MM1 is the envelope into which we put the blob to move it between the handset and the MMSC. The MM1 envelope includes information such as to whom the blob is to be sent, who will pay for getting it there, and the like.

An important property of the MM1 interface that differentiates it from all the other connections to the MMSC is that it is an air interface. This means simply that bits traveling over this connection go through the air during some part of their journey from the MMSC to the handset. Of course, that is the whole point of mobile telephony, but the fact that the bits go through the air puts some extremely challenging constraints on the MM1 interface.

This chapter and the next are primarily about MM1, moving multimedia messages from a handset to the MMSC, and moving messages from the MMSC to the handset. This interface is where the message meets the road, if you will, and is the key interface from your customers' point of view because it is the interface over which they get your application's messages. You have to understand this interface because it is at the end of the line. It is how you communicate with your customer and your customer communicates with you.

Before digging into the details of MM1 in the next chapter, let's complete our walk-through of the MMS architectural diagram. Since this book is about MMS application development, our interest is interfaces to the MMS architecture that can be used by application developers. One interface already noted is MM1. There are two others: MM3 and MM7. A chapter is dedicated to each of these interfaces later in the book. We will not cover the other internal MMS interfaces that the network operators use to talk to each other and to connect the MMS to other network functions, such as billing.

MMSC Connections to Other Computers

On the left-hand side of the MMS architectural diagram (Figure 2.1), you also see a box labeled "MMS User Agent B." This handset is connected to an MMSC that is different from the one to which "MMS User Agent A" is connected. Of course, the way it connects to this foreign MMSC is exactly the same as the way the "MMS User Agent A" connects to the local MMSC, through an MM1 interface. The architecture diagram shows the difference in how the local and foreign MMSCs talk to each other versus how they talk to handsets. MMSCs put the MMS bit blobs into MM4 envelopes, rather than MM1 envelopes, when they move the blobs between each other.

In theory and according to the defining specifications, the MM4, MM5, and MM6 interfaces are all inside the mobile telephone network and not visible to or usable by MMS application developers. In reality and in the near term, some MMSC manufacturers who haven't implemented the MM7 interface are making the MM4 interface do double duty as MM7, and thus presenting it—or a subset of it—to MMS value addded service providers. As this is an interim situation, MM4 will not be described in this book, either as an MMSC-to-MMSC connection, or as a proxy for the real MM7. Because they really are network internal, MM5 and MM6 will not be covered either.

Because this book is about developing MMS applications and not about setting up your own mobile network, we will concentrate on interfaces to the MMS system that application developers can see and use. The two other interfaces besides MM1 that MMS application developers can see and use are MM7 and MM3. They can be seen on the right-hand side of the MMS architecture diagram in both Figures 2.1 and 2.2.

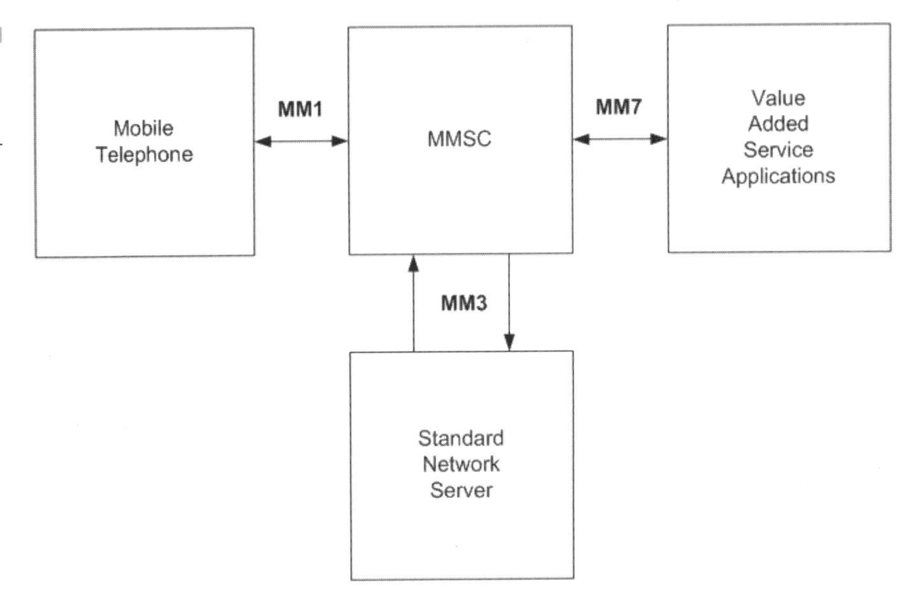

Figure 2.2 Application interfaces to MMS.

The MM7 interface is how the MMSC talks to value-added service providers (VASPs). A VASP could be an entity such as Yahoo that wants to let its customers create and send MMS messages from their Web portal, or it could be Sears wanting to send out appliance repair instructions as MMS messages. A private VASP could be a law enforcement agency that wants to send and receive mug shots as MMS messages.

If you are developing an MMS service, you are a candidate to be a VASP. You are probably planning (or at least hoping) to handle hundreds of MMS messages a day, not just one or two. For many reasons, not the least of which is plain old transmission speed, you do not want to have to send your messages over the MM1 air interface. You want a direct connection to the MMSC, and a richer language to provide your instructions to the network operator. This is the MM7 interface, to be described in Chapter 8.

Besides connecting to custom MMS services over the MM7 interface, an MMSC can connect to some predefined and standardized network services such as e-mail servers and various computer–telephone integration services such as fax-on-demand. When connecting to an existing network protocol, the MMSC uses the MM3 interface, rather than the MM7 interface. If the primary goal of your MMS application is to connect the corporate e-mail system to MMS, then you would use MM3 rather than MM7. MM3 will be described at length in Chapter 7.

MMS Media Elements

The atomic elements of an MMS message are pictures, sounds, and text or more properly speaking, digital representations of pictures, sounds, and text. Mobile devices available today support only a limited number of different representations of pictures and sounds, but the list of supported formats is expected to grow if MMS takes off as everybody in the mobile telephone business hopes it does. Table 2.1 lists what can be used today to build MMS messages.

TABLE 2.1
Today's MMS Media Formats

Content Type	Supported Formats
Pictures	JPEG, GIF87, GIF89a, WBMP
Audio	AMR (3GPP), EVRC (3GPP2), MIDI (both)
Free-running text	ASCII, UTF8, UTF16
Structured text	vCard, vCalendar

It must be stated that which media formats are supported and even what "support'" means are hotly debated topics. It is unfortunate that independent application developers and content providers are left out of these debates, at least to date. Virtually everyone will readily agree to the principle that for MMS to succeed, carriers and infrastructure providers must provide a stable and useful set of media formats. As yet, however, an agreement as to what that set consists of has not been reached.

The list in Table 2.1 is a based on the "MMS Conformance Document," published by a small group of handset manufacturers and MMSC suppliers. In theory, if MMS developers abide by the constraints of the conformance document, then the resulting MMS applications will play back successfully through and on all the MMS equipment manufactured by these companies. This is a really good idea because it sets a lower bar, or at least a common denominator, for your MMS messages.

Unfortunately, there is equipment on the market sold by some of these companies that doesn't handle MMS messages that conform to the conformance document, so MMS application developers cannot really count on anything yet. The only really reliable way to determine if your MMS will play back successfully on a particular handset is to buy the handset, load your MMS message on it, and see what happens.

Encoding pictures, sound, and text will be described in detail in Chapters 3 and 4.

MMS Message Assembly

The advantage MMS has over e-mail is that the pictures and sounds you send are not just attachments to a text message. The MMS developer can orchestrate very precisely how each element of a multimedia message is presented to the recipient.

For example, the developer could say, "Show Picture #1 for 10 seconds while playing Audio Track #1. Switch to Picture #2 after the 10 seconds is up, but continue to play Audio Track #1. At the 15-second mark, stop playing Audio Track #1 and start playing Audio Track #2. Continue this for another 10 seconds and stop both Picture #2 and Audio Track #2 and display Text #1 on the screen."

You can orchestrate your multimedia presentations with a language called the SMIL (pronounced "smile") or synchronized multimedia integration language. Figure 2.3 illustrates how SMIL ties together and coordinates the presentation of MMS media elements.

A SMIL program is itself simply a text file, like a Java program or a C program. Rather than talking about variables such as x and TOTAL, a SMIL program talks about media files such as Picture1.GIF and AuditTrack2.amr. Unlike a Java or C program, however, a SMIL program does not contain the values of what it is discussing. In a C program, when you say "x = 4," the value 4 is bound directly into the C program. A SMIL program talks about files containing pictures and sounds, but the files themselves are not inside the SMIL program; only the names of the files are inside.

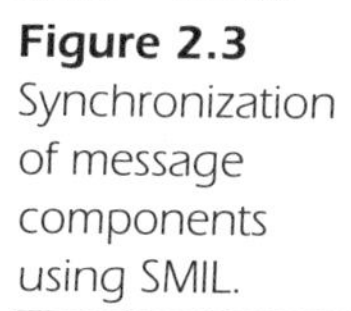

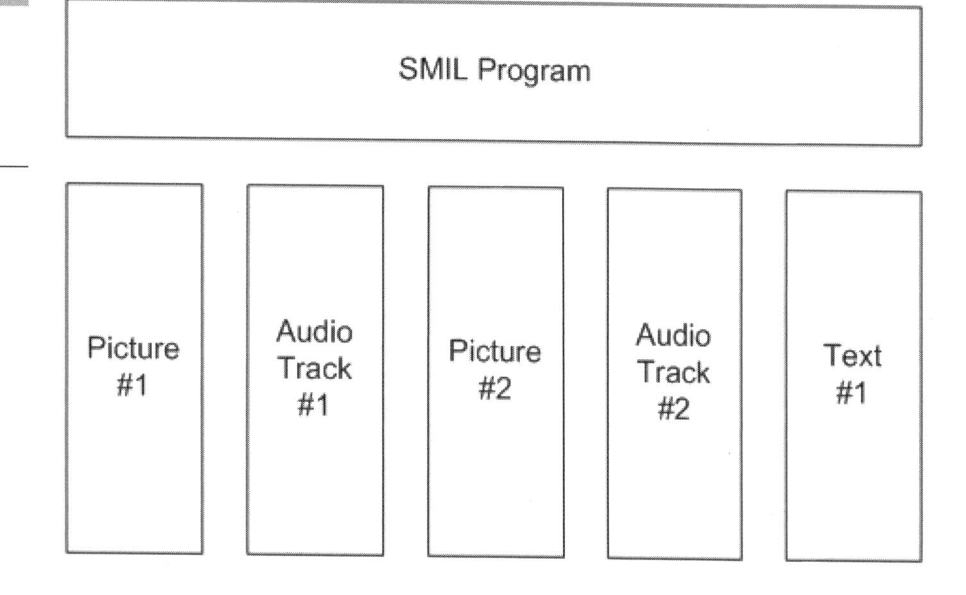

Figure 2.3 Synchronization of message components using SMIL.

To be sure the whole MMS message can be played when it reaches its destination, make sure that the files the SMIL program is talking about are packaged together with the program itself. MMS does this by using the very same packaging technique that e-mail uses, namely by multipurpose Internet mail extensions (MIME). Figure 2.4 illustrates how MIME packages the SMIL program and all related files into one message.

A pleasant side effect of the MIME packaging of MMS messages from an end user point of view is that you can send an MMS message via e-mail. The only downside is that the e-mail reading program opening the MMS won't replicate and play back the careful SMIL message orchestration described previously. All the files that make up the MMS message will just be displayed separately as attachments to the e-mail, at least as of early 2003. If MMS takes off, you can bet that e-mail programs will start to understand SMIL, and be able to play back MMS messages as faithfully as your mobile handset.

A more than just pleasant side effect of the MIME packaging of MMS messages, from a developer's point of view, is that all the tools and programs you have been using to build, process, and edit MIME e-mail can be put to work building, processing, and editing MMS messages.

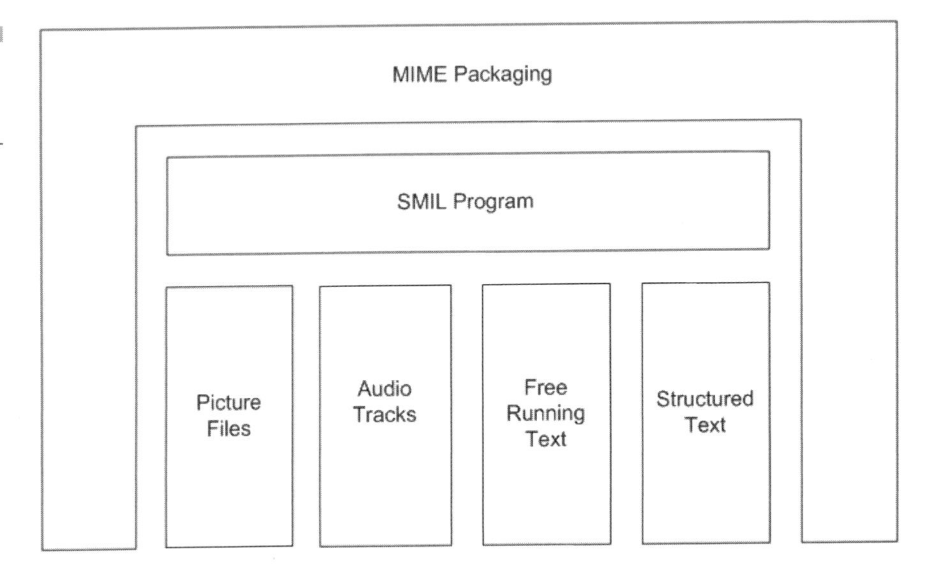

Figure 2.4 MIME encapsulation of an MMS message.

MIME packaging will be described in detail in Chapters 3 and 4.

MMS Message Passing

All digital communication is a matter of constantly putting something into an envelope, passing the envelope from point-to-point, and then taking whatever is in the envelope back out. Just like a postal envelope, a digital communication envelope contains lots of interesting information, such as to whom the message is going, where the message is coming from, what is inside the envelope, and how the transportation fees are to be paid.

If you could put two handsets side by side and move an MMS message from one to the other using the infrared port, the communication diagram would look something like Figure 2.5.

Figure 2.5 shows that the MMS message has been put into a MIME envelope and sent directly from one handset to the other. When the MIME envelope containing the MMS message gets to the receiving handset, it opens the envelope, takes out all the files, finds the SMIL program, and starts the playback. One of the bits of information that the MIME envelope contains is which of all the files it

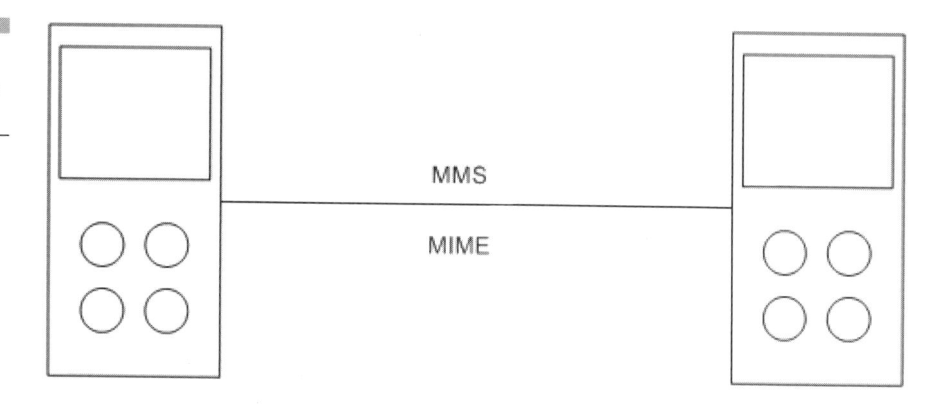

Figure 2.5 MMS message inside MIME envelope.

has encapsulated is the SMIL program. This is how the receiving handset knows how to start the playback.

MMS would not be very interesting—or very profitable to the network operators—if you could only move MMS messages directly from one handset to another. The MMS architectural diagram in Figure 2.1 shows what really happens—the MMS message moves from the sending handset to the MMSC using the MM1 interface, and then to the receiving handset, again using the MM1 interface. In other words, the real world picture is more like Figure 2.6.

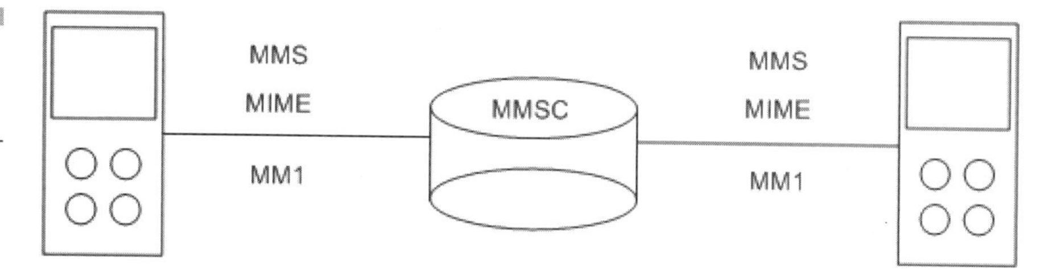

Figure 2.6 MIME package inside MM1 envelope.

We put the MIME envelope containing the MMS message into an MM1 envelope in the sending handset and pass the MM1 envelope to the MMSC after wireless session protocol (WSP) encoding it for efficient travel through the air. The MMSC examines the MM1 envelope for instructions and may send the envelope along untouched, or it may open the envelope and process the contents before passing them along. The next chapter is all about the MM1 envelope, but as you probably already guessed, one thing on the

MM1 envelope is the address of the MMS recipient. From 30,000 feet, the package that is passed between the MMSC and the mobile handset looks like Figure 2.7.

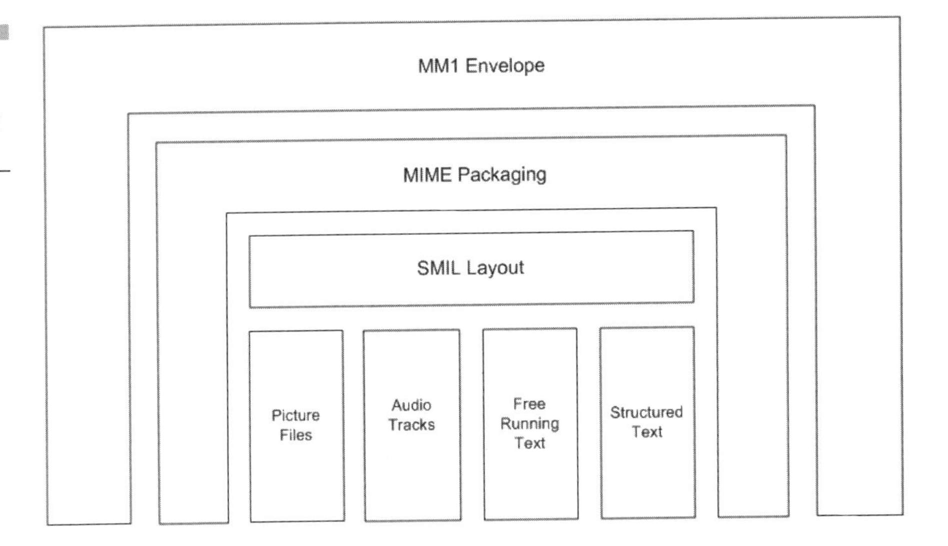

Figure 2.7 Structure of MMS message across MM1 interface.

Communication Bearers and Protocols

A detail not yet described regarding the direct transfer of an MMS from one handset to another using their infrared ports (illustrated in Figure 2.5) is how the two infrared ports talked to each other. You can bet they used some sort of envelope to move the MMS bits from one handset to the other.

Because this is a book about building MMS applications and not about wireless technology and network protocols, we will not go into the detail about the communication standards that are used to move MM1, MM3, or MM7 envelopes around. Nevertheless, it is of use to know a little about what is going on "under the covers."

MM3 and MM7 envelopes typically travel on the Internet protocol (IP) stack. Thus, for example, the MMSC would be connected to a VASP or another network server using plain old hyptertext transfer protocol (HTTP) over transmission control protocol (TCP).

MM1 envelopes are a little different because they go through the air, and we always have to worry about maximum efficiency when using a wireless link, particularly an expensive and heavily used wireless link such as the one to your mobile handset.

To realize this air interface efficiency, an additional node is inserted into the MMS architecture on the MM1 link. This node is a wireless application protocol (WAP) gateway. The responsibility of the WAP gateway is to efficiently encode the MM1 envelope and the MIME-encapsulated MMS message it contains, and get it onto the correct air link to the mobile handset.

The envelope that the WAP gateway uses to transmit this encoding is called the WSP. An example of WSP encoding in Chapter 3 gives you a sense of its efficiency. Thereafter, simply assume that the WAP gateway is silently doing its job, and an MM1 connection between the MMSC and the mobile handset or other end user device will be described. The position of the WAP gateway in the MMS architecture is illustrated in Figure 2.8.

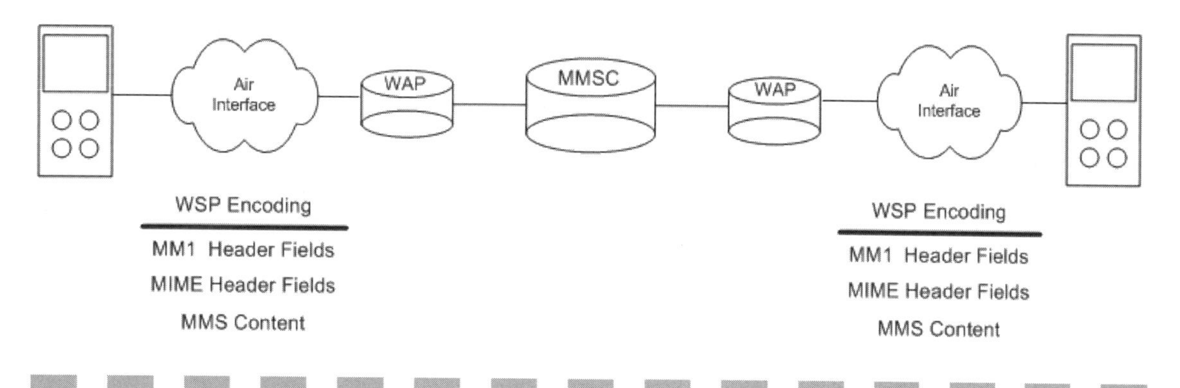

Figure 2.8 Network nodes in handset MMS communication.

The transmission of the MMS to the handset via the WAP gateway—which air interface bearer it uses—can vary. It depends on myriad factors including how the network operator has configured the system, the subscriber's account, the subscriber's current location, and the capabilities of the subscriber's handset, among others. For the purposes of designing a general-purpose MMS application, you can probably assume a GPRS connection because most operators talk about using one. Nevertheless, you should certainly check with your user population to see what is actually happening in the field in which you want to play.

The MMS Model of Operation

Like its predecessor, SMS, MMS is a store-and-forward system. For an application developer, this means you have no control over when your message arrives. What is new in MMS is that you can get slightly more feedback about when it did arrive and how it was handled by the intended recipient. Furthermore, since MMS messages are typically much larger than SMS, some additional control is also given to the recipient about how and when MMS messages are delivered, and what feedback is provided to the sender.

Figure 2.9 shows the basic MMS model of operation. The sender gives the MMS message to the MMSC, the MMSC sends a notification to the recipient that an MMS message is waiting, and finally, the recipient requests delivery of the message itself.

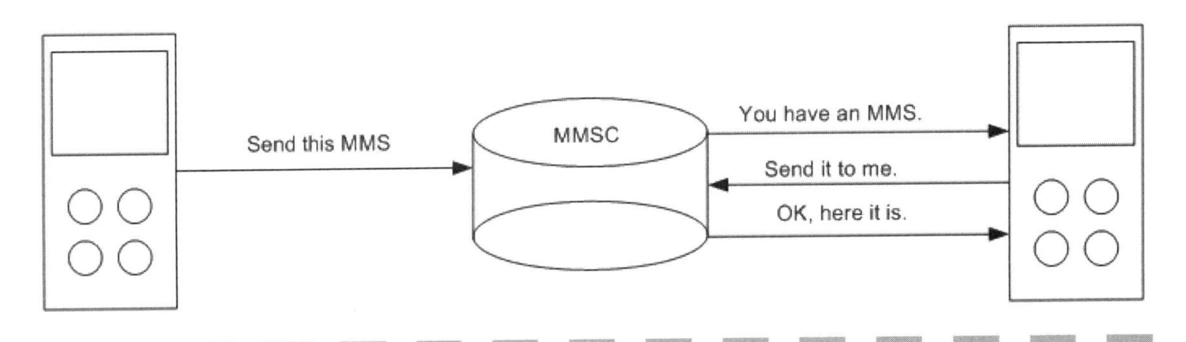

Figure 2.9 Basic MMS signalling.

Of course, there are many variations on this basic theme, and we will explore some of these as we dig into the details of the various interfaces to the MMSC.

What is useful to note with regard to the overall architecture of MMS compared to SMS is that: 1) the network operator will be holding much more data for 1,000 MMS messages than for 1,000 SMS messages; and 2) the input and output devices may be quite different, and it is possible that one or the other or both are not mobile devices. Both of these operational aspects of MMS have a profound impact on how an MMS system is both engineered and billed.

Another aspect of MMS that is quite different from SMS is the possibility that the network operator provides translation services. For example, if the operator knows that the recipient is using a handset that will handle JPEG images but not GIF images, the operator may translate the GIF images in an MMS going to the handset to JPEG images before sending them out over the air interface. Such a service will demand computational cycles that were never part of an SMS system design.

Summary

In this chapter, we described the overall architecture of the MMS. We briefly described each of the interfaces to the service that are of interest to the MMS application developer, together with the nature of the messages that circulate within the system.

What you have learned so far is:

- MMS is a store-and-forward messaging system.
- The three interfaces into the system of interest to application developers are MM1, MM3 and MM7.
- MMS messages are packages of individual media elements.
- The play back of the media elements in an MMS message is scripted using a language called SMIL.

The next two chapters describe in detail the most important interface in the system, MM1. This is the interface on which messages flow to and from the mobile handset.

CHAPTER 3

Structure of an MMS Message—The MM1 Interface

This chapter describes the process of building your own "Hello, world" example for multimedia messaging service (MMS). The idea is a one-time walk-through of all the bits-and-bytes details of an MMS message, so that you can see what is happening at this level. Knowing what is going on "under the hood" not only gives you an appreciation of the impact of your actions when using high-level development tools, but it also gives you the wherewithal to perform surgery at this level, should the need ever present itself. It is like being able to code in assembly language. You don't do it every day, but every once in a while you need a little patch of speed or access to something you can't reach with your high-level language, and you reach for your list of instruction codes.

Figure 3.1 is the storyboard of the MMS message we are going to build in this chapter. The stirring story begins with a display of an old-fashioned globe with a chorus of voices on the audio track going "Ooooooo." After 3 seconds, the display changes to a picture of the world, the display of the text "Hello, world," and thunderous applause on the audio track. Pretty exciting, huh?

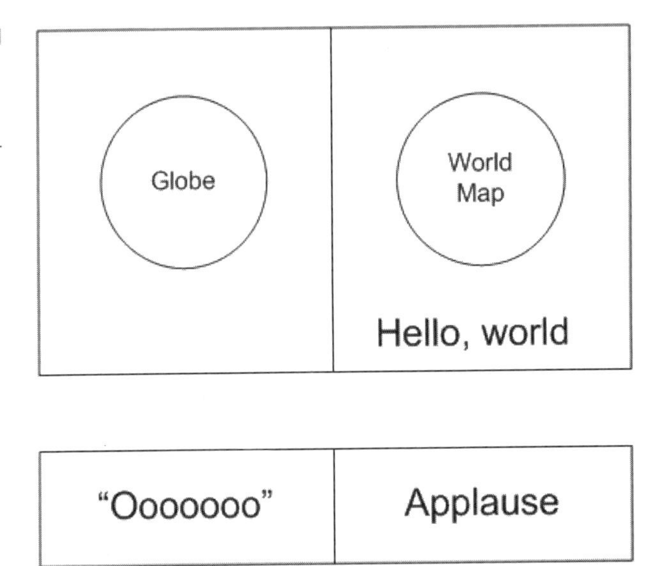

Figure 3.1 "Hello, world" as an MMS message.

Figure 3.2 is the structural diagram of the "Hello, world" message. We will build the message from the outside in, starting with the

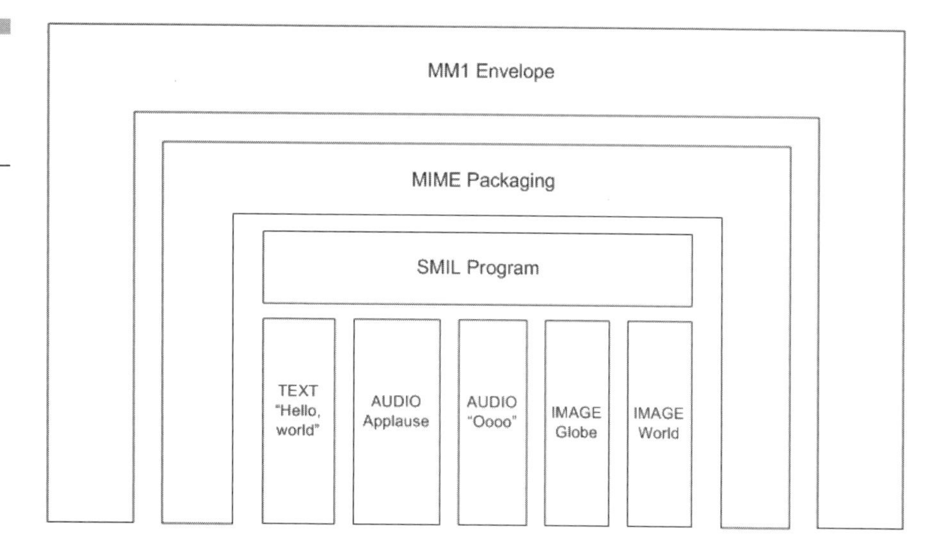

Figure 3.2 Structure of "Hello, world" MMS message.

MM1 envelope used to move the message between a handset and the Multimedia Messaging Service Center (MMSC), and finishing with the three media files that comprise the MMS message itself.

Request and Response Envelopes on the MM1 Interface

Communication between the MMSC and the handset on the MM1 interface is bidirectional—the handset can at any moment push an envelope to the MMSC, and the MMSC can at any moment push an envelope to the handset. Each such push is called a *request* and every request begets a *response* from the recipient.

Table 3.1 lists the requests that a handset can push to an MMSC on MM1, and Table 3.2 lists the requests that an MMSC can push to a handset on MM1.

Since each of these requests is paired with a reply, we will differentiate between the request envelope and the reply envelope by appending .REQ or .RES to the envelope name. For example, MM1_submit.REQ is the envelope that the handset sends to the MMSC to give it a new MMS message for someone. MM1_submit.RES is the response of the MMSC that could say, "Got it and it's on its way" or "Something is wrong with the MMS message. Send it again."

TABLE 3.1
MM1 Envelopes from the Handset to the MMSC

Envelope Name	Message Description
MM1_submit	"Here's an MMS I want to send to somebody."
MM1_retrieve	"Please transmit to me the MMS that has arrived."
MM1_forward	"Please forward the MMS that has arrived to somebody else."
MM1_acknowledgement	"Tell the sender that I have received the MMS they sent to me."
MM1_read_reply_recipient	"Tell the sender that I have played back the MMS they sent to me."

TABLE 3.2
MM1 Envelopes from the MMSC to the Handset

Envelope Name	Message Description
MM1_notification	"An MMS for you has arrived."
MM1_delivery_report	"The MMS you sent has been delivered."
MM1_read_reply_originator	"The MMS you sent has been read."

The format of an MM1 envelope plus its contents is always the same, whether it is a request envelope or a response envelope. If you are familiar with hypertext transfer protocol (HTTP), then you will feel comfortable because the format is exactly like the HTTP message format. First, there is a series of header fields, each one containing a header field name followed by a header field value (the information on the envelope). Following the header fields is the payload of the message (the content of the envelope). This basic format is illustrated in Figure 3.3.

The headers can be rendered in pure ASCII or in a highly compact binary form. For obvious reasons, the ASCII form is used when composing the headers and when discussing them, as we are doing now. The binary form is used when the envelope containing the headers is sent over the air. The binary form will be discussed later in this

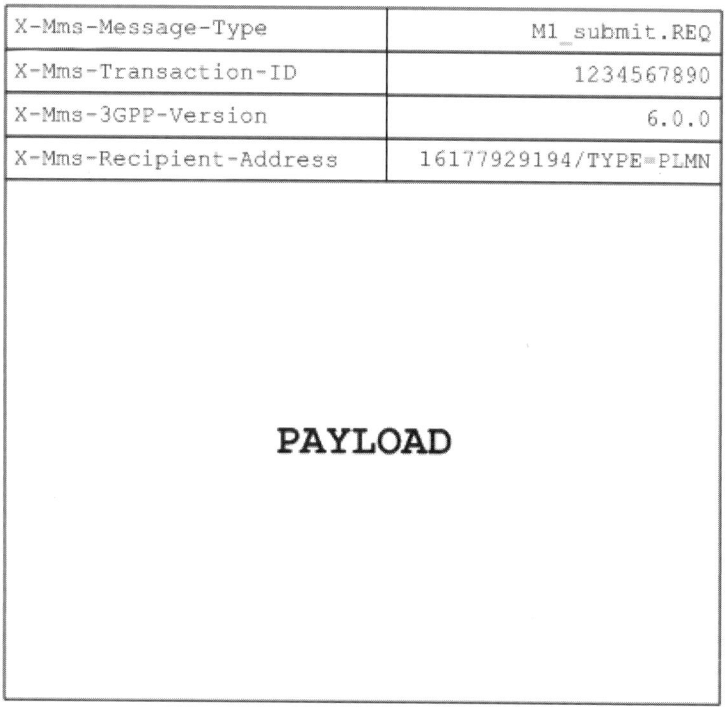

X-Mms-Message-Type	M1_submit.REQ
X-Mms-Transaction-ID	1234567890
X-Mms-3GPP-Version	6.0.0
X-Mms-Recipient-Address	16177929194/TYPE=PLMN

PAYLOAD

Figure 3.3 Typical headers of an MM1 envelope.

chapter. For now, we will use the ASCII form and write headers like this:

```
X-Mms-Message-Type: MM1_submit.REQ
X-Mms-Transaction-Id: 1234567890
X-Mms-3GPP-MMS-Version: 6.0.0
X-Mms-Recipient-Address: 16177929194/TYPE=PLMN
X-Mms-Subject: My First MMS
X-Mms-Sender-Address: sguthery/TYPE=RFC822@mobile-mind.com
X-Mms-Delivery-Report: Yes
```

In the ASCII form, the dividing line between the headers and the payload is two sequential carriage-return/linefeed pairs.

Every MM1 envelope, whether it is a request or a response, must start out with the same three header fields, namely:

```
X-Mms-Message-Type:
X-Mms-Transaction-Id:
X-Mms-3GPP-MMS-Version:
```

The Type header field, as you have already guessed, specifies which of the eight envelopes it is and whether it is a request (.REQ) or a response (.RES). The Transaction Id header identifies the envelope. For requests, it is an arbitrary character string, and for responses, it is the transaction ID of the request to which the response is a response. The Version header field specifies to which version of the 3GPP 23.140 specification the envelope complies. As of 2003, we are on Release 6 of this specification, so you would find 6.0.0 here.

The rest of the header fields depend on the type of the envelope. An MM1_submit.REQ envelope, for example, will have a set of additional header fields completely different from an MM1_notification.RES envelope.

The Basic MMS Model of Operation

The three primary envelopes on the mail interface are MM1_submit, MM1_notification, and MM1_retrieve. An originating handset uses an MM1_submit envelope to pass a new MMS message to the MMSC. The MMSC uses the MM1_notification envelope to tell a recipient handset that an MMS message for it has arrived. And finally, the recipient handset uses an MM1_retrieve envelope to tell the MMSC to send the MMS message itself.

You can see in these various envelopes the MMS version of the basic store-and-forward functionality of the MMSC, as compared to the same functionality in a Short Messaging Service Center (SMSC). An SMSC sends the text message to your handset preemptively, whether you want it or not. The MMSC is a little more formal. It tells you an MMS message has arrived and waits for you to request that it be sent before transmitting the message itself to your handset. There are a number of reasons for this.

First is the matter of size and format. Text messages are only 160 characters, and can easily be stored and displayed on any handset. MMS messages can be much larger and can include many different media formats, so you might want to switch from your text-only handset to your personal digital assistant (PDA) before you have the MMS message sent to you.

Second is the matter of expense. While all the details of MMS charging have not been worked out—and when they are, they will certainly vary from operator to operator—you may incur some sort of expense to have an MMS sent to your handset. The polite version of message download gives you the opportunity to decide whether to wait until evening or weekend rates come into effect before viewing the picture of your new grandchild having his or her diaper changed.

Last is the matter of forwarding. Either because you want to view the message on a personal computer (PC) rather than a handset, or because you want to save the message semipermanently, the "ask before sending" MMS download protocol lets you forward the message to a Web site or to your home computer via e-mail, rather than have it sent to your handset. If the message was sent to your handset preemptively, it is possible that you couldn't display it satisfactorily and that you have to pay the air interface charge again to send it back to the MMSC for forwarding on to your home computer.

Getting back to the various MM1 requests, two things probably occurred to you. First, some of the requests, in fact most of the requests, don't seem to contain a message. They are like postcards. The information in the headers is the message. This is true of both the MM1_notification.REQ and the MM1_retrieve.REQ envelopes.

Second, you asked yourself, where is the envelope that actually moves the MMS message to the receiving handset? As it turns out, the MMS message is moved to the receiving handset in response to the MM1_retrieve.REQ; that is, it is in MM1_retrieve.RES. This makes perfect sense. What better response to a request to send along an MMS message than the message itself, after all?

The MM1_submit Envelope

Of the eight messages on the MM1 interface, the most important one from the point of view of MMS application development is MM1_submit. If your application is a modest one—sending out 10 or

20 messages a day—or if you are just exploring the utility of MMS messaging for a bigger application, then using an MMS handset connected to a PC workstation or server and using the MM1_submit command is probably a good place to start. This avoids the cost and hassle of dealing with a value-added service provider (VASP) or directly with the network operator, and it gives you complete control of the messaging function. If nothing else, it teaches you what MMS can do, so that you can be an informed customer when you do want to scale up to an MM3 or MM7 connection.

Table 3.3 lists all the information that can go on an MM1_submit.RES envelope. You will use this envelope to beam your MMS message to the MMSC.

TABLE 3.3 Header Fields in an MM1 Submit Request Envelope

Field Name	Presence	Field Description
X-Mms-Message-Type	Mandatory	"MM1_submit.REQ"
X-Mms-Transaction-Id	Mandatory	Transaction identifier assigned to this message
X-Mms-3GPP-MMS-Version	Mandatory	"6.0.0"
X-Mms-Recipient-Address	Mandatory	Address of the recipient of the message
X-Mms-Sender-Address	Optional	The address of the originator of the message
X-Mms-Message-Class	Optional	For example, "personal," "advertisement," "information service," etc.
X-Mms-Date-And-Time	Optional	The date and time the message is submitted to the MMSC
X-Mms-Time-of-Expiry	Optional	The time at which the message should be deleted if it has not been delivered
X-Mms-Earliest-Delivery-Time	Optional	The earliest time at which an attempt should be made to deliver the message
X-Mms-Delivery-Report	Optional	"Yes," if the originator would like a delivery report
X-Mms-Reply-Charging	Optional	"Yes," if the originator is willing to pay for a reply to the message
X-Mms-Reply-Deadline	Optional	The latest time at which the originator is willing to pay for a reply

(continued on next page)

TABLE 3.3
Header Fields in an MM1 Submit Request Envelope (continued)

Field Name	Presence	Field Description
X-Mms-Reply-Charging-Size	Optional	The largest reply message measured in bytes for which the originator is willing to pay
X-Mms-Priority	Optional	The importance of the message
X-Mms-Sender-Visibility	Optional	"No," if the originator does not wish to reveal his or her address to the recipient
X-Mms-Read-Reply	Optional	"Yes," if the originator wishes to receive a read reply report
X-Mms-Subject	Optional	The Subject line of the message, which will be sent in the message notification
X-Mms-Reply-Charging-Id	Optional	In the case of a reply message, the message identifier of the message that is being replied to
X-Mms-Content-Type	Mandatory	MIME type of the MMS message body immediately following this header field

X-Mms-Message-Type

This is always set to `MM1_submit.REQ` in a submit request.

X-Mms-Transaction-Id

You can set this to any character string. Since you probably will not have more than one message outstanding with the MMSC, a sequential counter would work fine.

X-Mms-3GPP-MMS-Version

This is the version number of 3GPP TS 23.140 that describes this message. As of 2003, it is 6.0.0.

X-Mms-Recipient-Address

Currently, two forms of address are supported: telephone number and e-mail address. The format of a telephone number address (called an

MSISDN) is defined in excruciating detail by ITU-T E.164, and the format of an e-mail address is defined in IETF RFC 822. The address is followed by /TYPE= and then the type of the address. For example,

```
+16179461798/TYPE=PLMN
```

and

```
sguthery/TYPE=RFC822@mobile-mind.com
```

Other address types are envisioned in the future.

X-Mms-Sender-Address

Your address is in one of the preceding two formats. It is the address to which any reports or replies will be sent.

X-Mms-Message-Class

This can be "personal," "advertisement," "informational," or "auto," where "auto" does not mean it is about automobiles, but rather that the message has been automatically generated.

X-Mms-Date-And-Time

There are two types of time stamps: relative and absolute. Both are represented as 32-bit integers that are interpreted as number of seconds. Relative time is measured from the current time, and absolute time is measured from January 1, 1970 at 00:00:00 GMT.

The X-Mms-Date-And-Time field is the absolute time that the message was sent from the handset to the MMSC.

X-Mms-Time-of-Expiry

This time value allows the application to tell the network when it can delete the message if it has not been delivered. If it is a relative value, it is measured in seconds from X-Mms-Date-And-Time above.

X-Mms-Earliest-Delivery-Time

This field lets the application tell the network not to deliver the message before this point in time. If it is a relative value, it is measured in seconds from X-Mms-Date-And-Time above.

There are some obvious relationships between these times if they are expressed as absolute times. If the times are set incorrectly, the MMSC will probably reject your message.

X-Mms-Delivery-Report

This can be set to "Yes" or "No" and indicates if you would like to receive a report of the delivery of the message. This will probably cost extra. If you pick "Yes," the recipient can override your choice and refuse to return a delivery report. The default, if the header field is not included, is "No."

X-Mms-Reply-Charging

This field is set to "Yes" or "No" and specifies whether you are willing to pay for a reply. The default, if the header field is not included, is "No."

X-Mms-Reply-Deadline

If you are willing to pay for a reply, then you are only willing to pay if the reply is made before this time. If it is a relative value, it is measured in seconds from the preceding X-Mms-Date-And-Time.

X-Mms-Reply-Charging-Size

If you are willing to pay for a reply, then you are only willing to pay if the reply can be no larger than this number of bytes.

X-Mms-Priority

This field can be set to "Low," "Medium," or "High." What the setting means to the various network operators between you and your recipient, and the cost of the various settings, is still to be determined.

X-Mms-Sender-Visibility

Set this field to "No" if you do not want the recipient to know your address. The default is "Yes."

X-Mms-Read-Reply

This can be set to "Yes" or "No" and indicates if you would like to receive a report of the handling of the message by the recipient. This will also probably cost extra. Exactly like the delivery report, if you pick "Yes," the recipient can override your choice and refuse to return a read report. The default, if the header field is not included, is "No."

X-Mms-Subject

This arbitrary character string will be sent to recipients with the notification message so that they can determine how they would like to handle the message. E-mail rules apply.

X-Mms-Reply-Charging-Id

If this message is in reply to a previous message, then this is the X-Mms-Transaction-Id of that previous message. This lets the recipient of the reply associate the reply with the message that stimulated the reply.

X-Mms-Content-Type

This must always be the last header field, and it describes the content contained in the payload that immediately follows. For an MMS message, this field should be set to "application/vnd.wap.multipart.related." As you guessed, this simply means that the payload contains many related parts.

Following is the complete MM1 envelope for sending our "Hello, world" example to the MMSC:

```
X-Mms-Message-Type: MM1_submit.REQ
X-Mms-Transaction-Id: 1234-ABCD-5678
X-Mms-3GPP-MMS-Version: 6.0.0
X-Mms-Recipient-Address: 16177929194/TYPE=MSISDN
X-Mms-Subject: Hello, world
X-Mms-Sender-Address: sguthery/TYPE=RFC822@mobile-mind.com
X-Mms-Content-Type: application/vnd.wap.multipart.related
```

MIME Encapsulation

The payload of the MM1 envelope is itself another envelope. It is the MIME encapsulation of the six files that comprise the multimedia message, namely the SMIL file, the two audio track files, the two picture files, and the text file.

Eggs within eggs within eggs, a multipurpose Internet mail extension (MIME) encapsulation, is also many header fields followed by a payload. What is slightly different about MIME encapsulation, however, is that the payload is not an opaque, monolithic blob, but is a series of blobettes separated by a text string that is defined in the initial batch of headers.

Following is the next envelope in our example MMS message:

```
content-transfer-encoding: base64
content-type: multipart/related;
type="application/smil";start="<1000>";boundary="_----------
=_103140636136200"
MIME-version: 1.0
date: Sat, 7 Sep 2002 13:46:01 UT

This is a multi-part message in MIME format.

--_----------=_103140636136200
content-disposition: attachment;
filename="hello-world.smil"
content-transfer-encoding: 7bit
content-type: application/smil;
name="hello-world.smil"
content-id: <1000>
```

```
--_----------=_103140636136200
content-disposition: attachment; filename="globe.gif"
content-transfer-encoding: base64
content-type: image/gif; name="globe.gif"

--_----------=_103140636136200
content-disposition: attachment; filename="wooh.amr"
content-transfer-encoding: base64
content-type: audio/amr; name="wooh.amr"

--_----------=_103140636136200
content-disposition: attachment; filename="world.gif"
content-transfer-encoding: base64
content-type: image/gif; name="world.gif"

--_----------=_103140636136200
content-disposition: attachment; filename="applause.amr"
content-transfer-encoding: base64
content-type: audio/amr; name="applause.amr"

--_----------=_103140636136200
content-disposition: attachment; filename="Text0000.txt"
content-length: 12
content-transfer-encoding: 7bit
content-type: text/plain; name="Text0000.txt"

--_----------=_103140636136200--
```

There are a few things to note about the MIME envelope. First, look at the second header field that reads:

```
content-type: multipart/mixed; boundary="_----------
=_103140636136200"
```

This states that the MIME encapsulation is a collection of parts of different formats, and that the parts are separated by the text string `"_----------=_103140636136200"`. Looking down a little further, note that a file called "hello-world.smil" will be between the first and second appearances of this string, a file named "globe.gif" will be between the second and the third, and so forth.

Next, notice that there are not only header fields at the very beginning of the MIME envelope, but there are also header fields at the beginning of each part. The initial header fields describe the overall MIME envelope, and the header fields at the beginning of each part describe just that part. Thus, for example, the header fields for the second component say that the component is a GIF image that has been encoded using the base64 algorithm.

One final thing to note about the MIME envelope is the "start" header field in the collection of header fields at the beginning that describes the overall MIME message:

```
multipart/related; start="<1000>"; boundary="_----------
=_103140636136200"
```

This header field specifies which of the following components is the SMIL program that orchestrates the playback of the other parts. In this particular example, the "start" header field specifies that the SMIL program is the part with the identifier <1000>. Looking at the first part in the envelope, in the header fields describing this part is a header field called "content-id" that has the value <1000>. The mobile handset will look for a component part with a "content-id" header field whose value matches the value in the "start" header to find the SMIL program that orchestrates the MMS message.

The SMIL Program

As you can see in the MIME layout, the SMIL program is not itself an envelope with contents. It is just one of the parts of the MIME encapsulation, but as noted above, it is the part that is called out as the starting point of the message.

Rather than containing the files it uses, the SMIL program just refers to them. Following is the complete SMIL program for our "Hello, world" example:

```
<smil>
    <head>
        <layout><root-layout/>
```

```
            <region id="region1_1" top="0" left="0"
              height="100%" width="100%"/>
            <region id="region1_2" top="0" left="0"
              height="50%" width="100%"/>
            <region id="region2_2" top="50%" left="0"
              height="50%" width="100%"/>
        </layout>
    </head>
    <body>
        <par dur="3000ms">
            <img src="globe.gif" region="region1_1"/>
            <audio src="wooh.amr"/>
        </par>

        <par dur="3000ms">
            <img src="world.gif" region="region1_2"/>
            <text src="Text0000.txt" region="region2_2">
                <param name="foreground-color" value="#000000"/>
                <param name="textsize" value="large"/>
            </text>
            <audio src="applause.amr" end="2550ms"/>
        </par>
    </body>
</smil>
```

This is the exact text that would go between the line

```
content-id: <1000>
```

in the MIME encapsulation and the second appearance of the separator string:

```
--_----------=_103140636136200
```

The next chapter focuses on SMIL in detail, but we will quickly look at this SMIL program now to get a sense of how the language works.

The program begins by defining some regions of the screen. That is what appears between <layout> and </layout>. Later, the program will use these regions to place pictures and text on the screen. In our example, region1_1 is the whole screen, region2_1 is the top half of the screen, and region2_2 is the bottom half of the screen.

In the body section, note that the MMS message consists of two parts played sequentially, one after another. The first part appears between the first <par> ... </par> pair, and the second appears between the second <par> ... </par> pair.

The items between the <par> ... </par> pairs are played back in parallel (i.e., simultaneously). The first is played for 3,000 milliseconds or 3 seconds, and the second one is also played back for 3,000 milliseconds or 3 seconds.

Between the first <par> ... </par> pair is the display of globe.gif image in the whole screen (region1_1) and the simultaneous playback of the “Ooooooooooo” audio track.

Between the second <par> ... </par> pair is the display of the world.gif in the top half of the screen (region2_1) simultaneously, with the display of the contents of the Text0000.txt file in the bottom half of the screen (region2_2), and all of them simultaneously with the playback of the applause audio track. The attentive reader will notice that the applause track is only 2,550 milliseconds, long so there are 450 milliseconds of silence in which to ponder the deep meaning of the “Hello, world” MMS message before it disappears from the screen.

SMIL 2 MIME

The Perl script that follows takes as input a SMIL program and creates the complete MIME encapsulation for the MMS message that the SMIL program orchestrates.

The program uses two Perl modules: XML::Twig for scanning the SMIL program and MIME::Lite for building the MIME encapsulation. As each file used in the SMIL program is encountered, it is made into a part of the MIME encapsulation.

The Perl script just prints out the final MIME encapsulation. In actual production, it would add on the MM1 headers, optionally wireless session protocol (WSP) encode it, and then send it off to the MMSC either by using advanced technology (AT) commands and a mobile handset, or by appending some HTTP headers and using an HTTP connection to the MMSC.

```
use XML::Twig;
use MIME::Lite;

my ($smil_file) = @ARGV;

$mime = MIME::Lite->new(Type =>'multipart/related');

$mime->attr('Start' => '<1000>');
$mime->attr('Content-Type:' =>
application/vnd.mms.multipart.related');

$smil_part = MIME::Lite->new(
                Type            =>'application/smil',
                Path            =>$smil_file,
                Filename        =>$smil_file,
                Disposition     =>'attachment',
                Encoding        =>'7bit'
          );
$smil_part->attr('Content-ID' => '<1000>');

$mime->attach($smil_part);

my $twig = new XML::Twig (
                 TwigHandlers => {
                    'img'   => \&img,
                    'audio' => \&audio,
                    'text'  => \&text
                 }
               );

$twig->parsefile($smil_file);

print $mime->as_string;

sub img {
  my($twig, $img)= @_;
```

```
  my $image_file = $img->att('src');

  $image_file =~ /[^\.]\.(\w+)/;

  $mime->attach(
                Type            =>"image/$1",
                Path            =>$image_file,
                Filename        =>$image_file,
                Disposition     =>'attachment'
          );
}

sub audio {
  my($twig, $audio)= @_;
  my $audio_file = $audio->att('src');

  $audio_file =~ /[^\.]\.(\w+)/;

  $mime->attach(
                Type            =>"audio/$1",
                Path            =>$audio_file,
                Filename        =>$audio_file,
                Disposition     =>'attachment'
          );
}

sub text {
  my($twig, $text)= @_;
  my $text_file = $text->att('src');

  $mime->attach(
                Type            =>'text/plain',
                Path            =>$text_file,
                Filename        =>$text_file,
                Disposition     =>'attachment'
          );
}
```

MM1 Submit of the "Hello, World" MMS Message

All that remains now is to base64 encode the GIF and AMR files and to put them, along with the text file, between their respective separators in the MIME encapsulation. Appendix B is the entire "Hello, world" MMS message in ASCII form, ready to be sent off to the MMSC on the MM1 interface.

The Response of the MMSC

Every request begets a response, so the preceding submit request, which passes an MMS message from the handset to the MMSC, will result in a response from the MMSC back to the handset (see Table 3.4).

TABLE 3.4
Header Fields in an MM1 Submit Response Envelope

Field Name	Presence	Field Description
X-Mms-Message-Type	Mandatory	"MM1_ submit.RES"
X-Mms-Transaction-Id	Mandatory	Transaction identifier assigned to this message
X-Mms-3GPP-MMS-Version	Mandatory	"6.0.0"
X-Mms-Request-Status	Mandatory	The status of the message from the point of view of the MMSC
X-Mms-Request-Status-Text	Optional	Text describing the status above that can be displayed to the user
X-Mms-Message-Id	Mandatory	The message identifier assigned by the MMSC to the MMS message in the body

The values of the first three header fields echo the values in these fields in the submit request. The other three fields are as follows:

X-Mms-Request-Status

This is an error code that specifies whether the MMSC likes what it received, and if it did, what it did with it. The possible codes are given in Table 3.5. These are the status codes for any response envelope, not just for the response to an MM1_submit.

TABLE 3.5
MM1 Request Status Codes

Status Code	Meaning
0	OK
1	Unspecified error
2	Service denied
3	Message format corrupt
4	Sending address unresolved
5	Message not found
6	Network problem
7	Content not accepted
8	Unsupported message

"Content not accepted" does not mean that the MMSC does not approve of a picture you sent. It means that it cannot handle one of the MIME media formats that you used.

X-Mms-Request-Status-Text

This is text generated by the MMSC that reflects the preceding status code. The text itself is not standardized and may vary from MMSC vendor to MMSC vendor, but it will always give the user a sense of what has happened to the message and why. In the usual case where everything goes well, the status text will just be "Message sent" or something similar.

X-Mms-Message-Id

This is an identifier that the MMSC assigns to the message. The MMSC will reference this identifier when passing back additional information about the message, such as when it was finally delivered and when it was read by the recipient.

Assuming our MMSC likes our "Hello, world" message, the reply back will be the following:

```
X-Mms-Message-Type: MM1_submit.RES
X-Mms-Transaction-Id: 1234-ABCD-5678
X-Mms-3GPP-MMS-Version: 6.0.0
X-Mms-Request-Status: 0
X-Mms-Message-Id: 9AvY38M1jPyo73W
```

Notification of the Arrival of the "Hello, World" MMS Message

The delivery of an MMS message to a handset on the MM1 interface starts with a message pushed from the recipient's MMSC to the recipient's handset notifying the recipient that an MMS message has arrived for him and it is available for download. It ends with the optional return of a message to the sender that the notification message has been read by the recipient. To follow the discussion, take off your sender's hat and put on your recipient's hat.

The push message could be a plain old short messaging service (SMS) message or a wireless application protocol (WAP) push, so that the notification ends up in your handset's MMS in-box, rather than the SMS in-box.

The contents of the notification message is an MM1 envelope with no payload (see Table 3.6).

TABLE 3.6

Header Fields in an MM1 Notification Request Envelope

Field Name	Presence	Field Description
X-Mms-Message-Type	Mandatory	"MM1_notify.REQ"
X-Mms-Transaction-Id	Mandatory	Transaction identifier assigned to this message
X-Mms-3GPP-MMS-Version	Mandatory	"6.0.0"
X-Mms-Message-Class	Mandatory	For example, "personal," "advertisement," "information service," etc.
X-Mms-Message-Size	Mandatory	Total number of bytes in the body of the MMS
X-Mms-Time-of-Expiry	Mandatory	Time at which the MMSC will delete the MMS if action is not taken
X-Mms-Message-Reference	Mandatory	Uniform resource identifier (URI) for the message that can be used to retrieve it from the MMSC
X-Mms-Subject	Optional	Subject line assigned by the sender
X-Mms-Priority	Optional	The importance of the message
X-Mms-Sender-Address	Optional	Address of the sender of the message
X-Mms-Stored	Optional	The message has been automatically put into the subscriber's MMBox
X-Mms-Delivery-Report	Optional	"Yes," if the sender has asked for a delivery report
X-Mms-Reply-Charging	Optional	"Yes," if the sender is willing to pay for a reply
X-Mms-Reply-Deadline	Optional	Deadline by which the sender is willing to pay for a reply
X-Mms-Charging-Size	Optional	Maximum size of the reply for which the sender is willing to pay
X-Mms-Charging-Id	Optional	If this is a reply message, this is the X-Mms-Message-Id of the original message
X-Mms-Distribution-Indicator	Optional	If set to "false," the VASP has indicated that content of the MM is not intended for redistribution If set to "true," the VASP has indicated that content of the MM can be redistributed

The notification message received by your handset for the "Hello, world" MMS might look something like this:

```
X-Mms-Message-Type: MM1_notification.REQ
X-Mms-Transaction-Id: 1234-ABCD-5678
X-Mms-3GPP-MMS-Version: 6.0.0
X-Mms-Message-Class: Personal
X-Mms-Message-Size: 88640
X-Mms-Time-of-Expiry: 10000
X-Mms-Subject: Hello, world
X-Mms-Message-Reference:
http://123.45.67.890/16177929194/7YteZ3u1nBc
X-Mms-Sender-Address: sguthery@mobile-mind.com
```

Notice that the MMSC is obliged to pass along some important bits of data that the recipient can use to decide whether the MMS should be downloaded to the handset and, if so, when the download should take place.

The first most important header field value for this purpose is the size of the MMS. If the MMS is big, and you are just getting into your car, you probably do not want to download it now. In fact, if it is really big, you may want to forward it to your desktop, rather than take delivery on your handset.

The next bit of critical information is the expiration time. The receiving MMSC is not obliged to hang onto incoming MMSs forever and may, in fact, implement a dynamic strategy, based on the size of the MMS. Another possibility is that you get charged for space on the MMSC taken up by your incoming MMSs. Either way, the expiration time says when the MMSC will delete the message if it has not been picked up or otherwise handled.

The expiration time can be absolute or relative. If absolute, it will be a time such as September 23, 2003 at 6:30 A.M. The exact format of absolute dates is given in the canonical HTTP specification, RFC 2616. If relative, it will be a number of seconds from the time the notification message was sent. Your handset will probably translate this into something more compelling like days, hours, and minutes.

The message reference in the notification tells your handset where to get the MMS message itself. Whether you can hit this URL from your desktop is up to the network operator.

Since this request was initiated by the MMSC, you (or more exactly your handset) are obliged to send a response. Remember, every request begets a response. Table 3.7 shows the header fields in this response. Note that there is no payload.

TABLE 3.7

Header Fields in an MM1 Notification Response Envelope

Field Name	Presence	Field Description
X-Mms-Message-Type	Mandatory	"MM1_notify.RES"
X-Mms-Transaction-Id	Mandatory	Transaction identifier assigned to this message
X-Mms-3GPP-MMS-Version	Mandatory	"6.0.0"
X-Mms-Message-Status	Optional	The status of the reception of the notification message; almost certainly 0
X-Mms-Report-Allowed	Optional	"Yes" if a requested delivery report can be sent by the MMSC

When you send an MMS to somebody, you can request that either one or both delivery reports be returned to you. The first report is generated when the recipient receives a notification message that an MMS is available. The second report is generated when the recipient actually fetches the MMS message itself.

If the message came in over the MM1 interface, then the recipient of the MMS can block the sending of these reports if the message was sent from another mobile phone. That is what the X-Mms-Report-Allowed means. If it is set to 1, the recipient is saying to the MMSC, "If the sender asked for a delivery report, it is OK to send it." If it is set to 0, the recipient is saying, "If the sender asked for a delivery report, ignore the request and do not send it."

Note carefully that the ability of an MMS recipient to block delivery reports pertains only to MMS messages coming from other mobile phones; that is, those sent into the MMS system over the MM1 interface. If the MMS message was entered into the MMS system over the VASP interface, MM7, then the recipient *cannot* block the return of the delivery reports. The MMS system sends back the report, whether the recipient likes it or not. Whether the recipient is notified that a report is being returned is still uncertain.

From an application developer's point of view, what is important is that if you need to get delivery reports for whatever reason—billing, for example—you have to send your MMS messages into the system over an MM7 interface.

Also notice in passing that the response to the notification request does not tell the MMC what to do with the MMS.

Delivery of the "Hello, World" MMS Message

Finally, the big moment has arrived. You are on the edge of your seat. What *is* this strange MMS with the subject "Hello, world"? Let's go get it and see.

To retrieve the message, you push an MM1 envelope with no payload back to the MMSC holding the message. The header fields possible in this envelope are given in Table 3.8.

TABLE 3.8
Header Fields in an MM1 Retrieval Request Envelope

Field Name	Presence	Field Description
X-Mms-Message-Type	Mandatory	"MM1_retrieve.REQ"
X-Mms-Transaction-Id	Mandatory	Transaction identifier assigned to this message
X-Mms-3GPP-MMS-Version	Mandatory	"6.0.0"
X-Mms-Message-Reference	Mandatory	URI of the message to be retrieved that was contained in the notification request

Simplicity itself. All fields are mandatory and the request simply says, "Download this message now, please." Following is what our example might look like:

```
X-Mms-Message-Type: MM1_retrieve.REQ
X-Mms-Transaction-Id: 1234-ABCD-5678
X-Mms-3GPP-MMS-Version: 6.0.0
X-Mms-Message-Reference:
http://123.45.67.890/16177929194/7YteZ3u1nBc
```

The mandated response to this request is the MMS message. Table 3.9 lists all the MM1 header fields that might precede the MIME encapsulated MMS message. As with the MM1 envelope used to hand the message to the MMSC in the first place, the X-Mms-Content header field appears last in the list of header fields and just before the MIME-encapsulated MMS message itself.

TABLE 3.9

Header Fields in an MM1 Retrieval Response Envelope

Field Name	Presence	Field Description
X-Mms-Message-Type	Mandatory	"MM1_retrieve.RES"
X-Mms-Transaction-Id	Mandatory	Transaction identifier assigned to this message
X-Mms-3GPP-MMS-Version	Mandatory	"6.0.0"
X-Mms-Originator-Message-Id	Mandatory	The message identifier of the MMS messag
X-Mms-Sender-Address	Conditional	Address of the originator of the MMS, unless he indicated that his address was to be hidden
X-Mms-Content-Type	Mandatory	The type of content contained in the MMS message (e.g., audio, mixed, etc.)
X-Mms-Recipient-Address	Optional	Address of the recipient(s) of the message; includes current addressee, but may include others if the message was multiply addressed
X-Mms-Message-Class	Optional	The class of the message (e.g., "personal," "advertisement," "information service," etc.)
X-Mms-Date-And-Time	Mandatory	The date and time the message was sent or forwarded to the current recipient
X-Mms-Delivery-Report	Optional	"Yes," if a delivery report is requested by the sender
X-Mms-Priority	Conditional	The importance of the message, as assigned by the originator of the message
X-Mms-Read-Reply	Conditional	"Yes," if a read report is requested by the sender
X-Mms-Subject	Conditional	The Subject line assigned by the originator of the message

(continued on next page)

TABLE 3.9 Header Fields in an MM1 Retrieval Response Envelope (continued)

Field Name	Presence	Field Description
X-Mms-State	Optional	Describes the disposition of the MMS in the subscriber's MMBox
X-Mms-Flags	Optional	More information about the status of an MMS stored in the subscriber's MMBox
X-Mms-Status	Optional	The status of the retrieve request
X-Mms-Status-Text	Optional	Test that qualifies or explains the above status code
X-Mms-Reply-Charging	Optional	"Yes," if the sender is willing to pay for a reply to this message
X-Mms-Reply-Charging-ID	Optional	If this message is itself a reply, then this is the message identifier of the original message
X-Mms-Reply-Deadline	Optional	The latest time that the sender is willing to pay for a reply to the message
X-Mms-Reply-Charging-Size	Optional	The maximum size of the reply message that the sender is willing to pay for
X-Mms-Previously-Sent-By	Optional	The list of addresses of people who have forwarded this message. Includes the original sender
X-Mms-Previously-sent-date-and-time	Optional	The dates and times of the forwarding events
X-Mms-Message-Distribution-Indicator	Optional	If set to "false," the VASP has indicated that content of the MM is not intended for redistribution If set to "true," the VASP has indicated that content of the MM can be redistributed
X-Mms-Content	Conditional	The content of the message itself follows this header element

A lot is happening here, but of particular note from an application developer's point of view is the X-Mms-Message-Distribution-Indicator bit. If it is set, then the handset will block resending of the MMS. This bit is the tippy top of the digital rights management iceberg that is floating in the MMS sea.

The protection of content in MMS messages will be described later in the book, but it is important to note that this bit can only be set over the VASP or MM7 interface. It cannot be set over the mobile phone or MM1 interface. If you want to protect your content, then you have to use the VASP interface. (It would seem that there might be times when mobile phone subscribers would want to set this bit on pictures they take with their mobile phone cameras and send on MM1, but that is another story.)

The good news is that the MMSC can give you a lot of information about the MMS message, in addition to the message itself. But, of course, the beef is the MMS message itself. Without the "Oooooooooooooo" background for the first image and the applause background for the second image, the received message is shown in Figure 3.4 as captured by the T68i handset simulator that is part of the Sony Ericsson MMS Composer (http://www.ericsson.com/mobilityworld/sub/open/technologies/messaging/tools/mms_composer).

Figure 3.4 "Hello, world" MMS message.

Other MM1 Envelopes

There are five other envelopes on the MM1 interface between the MMSC and the handset, besides the three primary ones that we used to move the MMS message through the system. These five other MM1 envelopes are summarized in Table 3.10.

TABLE 3. 10
Ancillary Envelopes on the MM1 Interface

Envelope Name	Direction	Message Description
MM1_forward	Recipient Handset to Recipient MMSC	"Please forward the MMS that has arrived to here."
MM1_acknowledgment	Recipient Handset to Recipient MMSC	"Tell the sender that I have received the MMS they sent to me."
MM1_delivery_report	Sending MMSC to Sending Handset	"The MMS you sent has been delivered to the recipient."
MM1_read_reply_recipient	Recipient Handset to Recipient MMSC	"Tell the sender that I have read the MMS they sent to me."
MM1_read_reply_originator	Sending MMSC to Sending Handset	"The MMS you sent has been read by the recipient."

In response to an MM1_notification, the MM1_forward envelope lets the receiving handset send a received MMS to somewhere else besides the handset.

The next four MM1 envelopes are two pairs. The first pair lets the receiving handset notify the sending handset that the MMS message arrived, and the second pair lets the receiving handset notify the sending handset that the MMS message has been read.

A Cautionary Word about Transcoding and Content Adaptation

Transcoding and content adaptation are technical jargon for the automatic reformatting of an MMS message by the MMSC, or by a transcoding engine connected to the MMSC. The announced purpose of transcoding is to make the MMS look as good as possible on the handset to which it is being sent. The assumption is that the network operator knows this information and the sender of the MMS does not, and that the operator is doing this as a service to his or her customer, the recipient.

Other less widely discussed purposes of transcoding by the network operator include charging extra for the use of high-quality handsets, generating revenue by adding advertising onto MMS messages, censoring of MMS content, and checking for digital rights violations. Performing these various transcoding functions is seen as a big business opportunity both by the operators and by those who sell servers and services to the operators.

The aspect of transcoding that is of primary concern to MMS application developers is reformatting of the MMS with respect to the recipient's current state, such as the handset they are using. This aspect of transcoding shifts at least some control of the presentation, impact, and user perception of the MMS from the application developer to the network operator. What is distressing is that you, the application developer, do not know and cannot find out what transformations will be made to your application, or under what circumstances they will be applied.

If you know exactly what handset your customer is carrying and you take pains to craft the MMS so that it creates the impression you want, you are not going to take kindly to having your work undone by a nameless, faceless programmer working for an MMSC vendor somewhere, or the policy of a network operator who thinks of the recipient as his customer, not yours.

If you have ever worked with format translation programs, particularly audio translators, you know that the quality of the result depends on more than just a few fixed parameters of the input. And yet a few fixed parameters are all the typical MMSC or transcoding engine will have the time or the resources to consider in any high-volume, production environment. And if the transcoding also involves transformations beyond a simple format change, the number of menu choices in Photoshop or GoldWave should convince you that applying these transformations blindly will stand as much chance of making things worse as they do of making things better.

But even when the transformation is as simple as resizing the images in the MMS to fit the recipient's screen, different strategies yield radically different results. Are the images cropped to fit? If so, is any important part of your presentation lost? Are the images resized to fit? If so, is the aspect ratio maintained, or do people become short and fat on landscape screens, but tall and thin on portrait screens?

There are a number of factors that work against the hidden and gratuitous reformatting of MMS messages by the network operator over and above the added cost and unbillable network traffic of doing it.

First, VASPs have the ability to suppress reformatting on MMS messages that are submitted over MM7 interface. If operator reformatting becomes a problem for MMS application developers, they can either become a VASP or partner with a VASP and set the bit that turns reformatting off. Of course, the VASP will add a charge on top of the operator's charge to handle the MMS. And sooner or later, it will occur to many application developers that they are paying extra to ensure that a service is not rendered.

Next, unless the number of different kinds of handsets settles down, the network operators and their MMSC vendors are faced with the costly and ongoing administrative task of keeping the knowledge base of handset properties and capabilities of all the network MMSCs synchronized and up-to-date. If history is any guide, the handset manufacturers are not likely to commodify themselves and start producing look-alike handsets. Nor are the network operators likely to stop ordering handsets of their own special design. If anything, MMS and other multimedia mobile services will amplify this trend.

The third scenario that might mitigate against transcoding is a future wherein the bulk of MMS messages will be created by content providers who either supply the handset or know themselves what handset their customer is packing. In this case, the content provider or a service used by the content provider will do the content adaptation to maximize the satisfaction of the recipient. If network operators transcode MMS messages coming from content providers, they are in fact interfering with the content provider's business and tending to remove competitive advantages based on message construction and delivery that one content provider may create over another.

Next, current thinking is that MMSs will be billed by size. Transcoding will alter size and, therefore, neither the sender nor the recipient will have any way of estimating the cost of sending or receiving an MMS before the fact. If an MMS image that uses a minimalist palette of only 256 colors is puffed up to an image that could use a palette of 65,535 colors because that is the capability of the receiving handset, the cost will go up. The image still uses only 256 colors, but your recipient will pay more because it could have used 65,535.

If, on the other hand, the transcoding engine decides to alter any of the colors in the palette—for example, dithering in some new colors when the receiving device has a bigger palette than used by the image, or combining colors when the receiving device has a smaller palette—then the color of a logo or other registered service mark in the image could be changed, and this might not be what the owner of the service mark thinks is within the rights and responsibilities of the network operator.

Fifth is the issue of what transcoding will do to digital rights management techniques, such as embedded watermarks that application developers and content providers have applied to the content of an MMS. If transcoding weakens or removes these digital protection rights, then the owner of the rights might have an issue with the entity that weakened or removed them.

The same reasoning applies to MMS messages that have been secured using fingerprints or digital signatures. Any transcoding will invalidate these protections. With respect to MMS media fingerprints, such as those promoted by Beep Science, it is unlikely that the fingerprint database will contain the fingerprint for every possible transcoding of the media element. As soon as transcoding creates a version of the element whose fingerprint is not on file, the forward progress of an MMS message that is in fact legitimate will cease.

With respect to digital signatures, the transcoding engine will not be in possession of the private key that created the signature, and the recipient will reject the message because the signature will not validate.

A transcoding engine will have a very hard task of figuring out if the components of an MMS message contain either watermarks, digital signatures, or fingerprints on file somewhere and, as a result, avoiding doing damage in these cases.

Finally, transcoding could actually cause bad things to happen. Suppose that in the process of transcoding, the location of objects on the screen changed so that they were no longer aligned with the regions defined by the <a> and <area> hot buttons in the SMIL script. Then recipients could think they were picking one alternative based on the location of the objects, but would actually be picking another based on the positioning of the hot buttons. "I said to click STOP and it behaved as if I clicked GO."

In spite of all this, many companies still want to get into the business of selling transcoding and content adaptation servers, as Table 3.11 shows.

TABLE 3.11 MMS Transcoding and Content Adaptation Products

Company	URL
PictureIQ	www.pictureiq.com
ConVisual	www.convisual.com
Mobixell	www.mobixell.com
LightSurf	www.lightsurf.com
Comverse	www.comverse.com

Historically, telecommunications companies have carefully avoided becoming in any way involved in the meaning or content of the communications they carry. Transcoding goes beyond simply knowing what is in the message. It makes the network operator a co-author of the message and, therefore, party to any consequences of the message.

At the end of the day, operators may elect to leave the MMS envelope closed because opening it only provides cost and trouble. Maybe transcoding was just a cover for adding advertisements to the messages all along, and they will discover they can do this without mangling the message.

Summary

In this chapter, we walked through the transmission of a complete MMS message called "Hello, world" from an MMS handset to the MMSC, and from the MMSC to a handset. We described the various envelopes on the MM1 interface that are used to communicate between the sending handset and the sending MMSC, as well as the receiving handset and the receiving MMSC. We also covered how MIME is used to encapsulate the components of an MMS message into a single multipart message that is sent through the multimedia messaging system.

In the following chapters, we suppress the details of both the MM1 envelopes and the MIME encoding, except when some particular feature or property has to be called out. The following chapter focuses on the construction of MMS messages and the use of the SMIL language.

CHAPTER 4

MMS Development Tools and Resources

Let's set aside for the moment all of the scaffolding and machinery that is used to move multimedia messaging service (MMS) messages about the network, and focus on the creation and testing of the MMS messages themselves.

An MMS message is a layout in time and space of visual, audio, and text elements. The language used to describe the layout is the synchronized multimedia integration language (SMIL). The general construction of an MMS message using SMIL is illustrated in Figure 4.1.

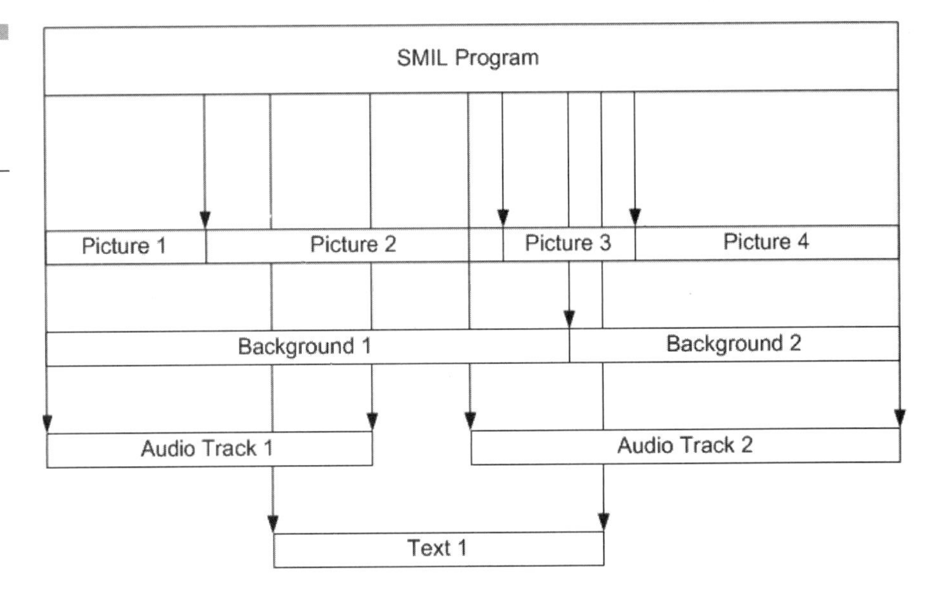

Figure 4.1 Uisng SMIL to synchronize an MMS message.

For each media element—text, graphics, and audio—the SMIL coordination script dictates when the element should begin and when it should end. In addition, for visual and text elements, the SMIL script defines where on the screen the elements should be appear.

The SMIL used for MMS messages is a subset of the Full SMIL language described in W3C specifications for SMIL. This subset is called 3GPP SMIL. 3GPP SMIL is a proper superset of the W3C's Basic SMIL, and a proper subset of the W3C's Full SMIL. In other words, it is Basic SMIL with some added bits and pieces of Full SMIL. The good news is that nothing in 3GPP SMIL is unique to MMS messages. 3GPP SMIL is described in 3GPP TS 26.234, "Transparent end-to-end packet switched streaming service (PSS): Protocols and codecs."

The bad news is that 3GPP SMIL is only found in the standards document, not in the field. In the field, four other flavors of MMS SMIL are in play: Basic SMIL, Conformance SMIL, Global System for Mobile Communication (GSM) Association SMIL, and whatever the handset you are working with really understands.

Note that the 3GPP2 and code division multiple access (CDMA) networks are also rolling out MMS-like functionality. Their MMS standards process is somewhat behind the 3GPP standards process primarily because a number of proprietary picture and multimedia messaging systems are in the field already. None of these messaging systems use SMIL, so at this writing, it is not clear how or when interoperability between 3GPP and 3GPP2 MMS will be achieved.

Conformance SMIL is defined in the MMS Conformance Specification, written by the MMS Interoperability Group consisting of the Computer Management Group, Comverse, Ericsson, Logica, Motorola, Nokia, Siemens, and Sony Ericsson. Conformance SMIL is a minimalist subset of Basic SMIL and not really suitable for much more than building mobile greeting cards. As noted in previous chapters, building an MMS in conformance with the Conformance Document does not ensure that the application works consistently on the equipment of all these manufacturers. For example, the Nokia 7650 is advertised as being an MMS handset, but it does not process the SMIL in an MMS message. The Motorola T720i only plays back the first three frames of an MMS message.

GSM Association SMIL is a variant of SMIL being put forward by the GSM network operators. It is Conformance SMIL with some wholly new constructs added that are not found in anybody else's SMIL. Why network operators want to be in the scripting language design business is anyone's guess.

The sixth and final version of MMS SMIL, and the only one that really counts, is the SMIL that is implemented in your recipient's handset. Unfortunately, you are not allowed to know this one. The handset manufacturer defines this version of SMIL, although it could also be affected by content adaptation services deployed by the network operator. It may be Conformance SMIL. It may be Basic SMIL. And it could even contain elements that are not in the Full SMIL document at all.

Figure 4.2 is a graphic representation of the variants of SMIL that are discussed when working with MMS messages.

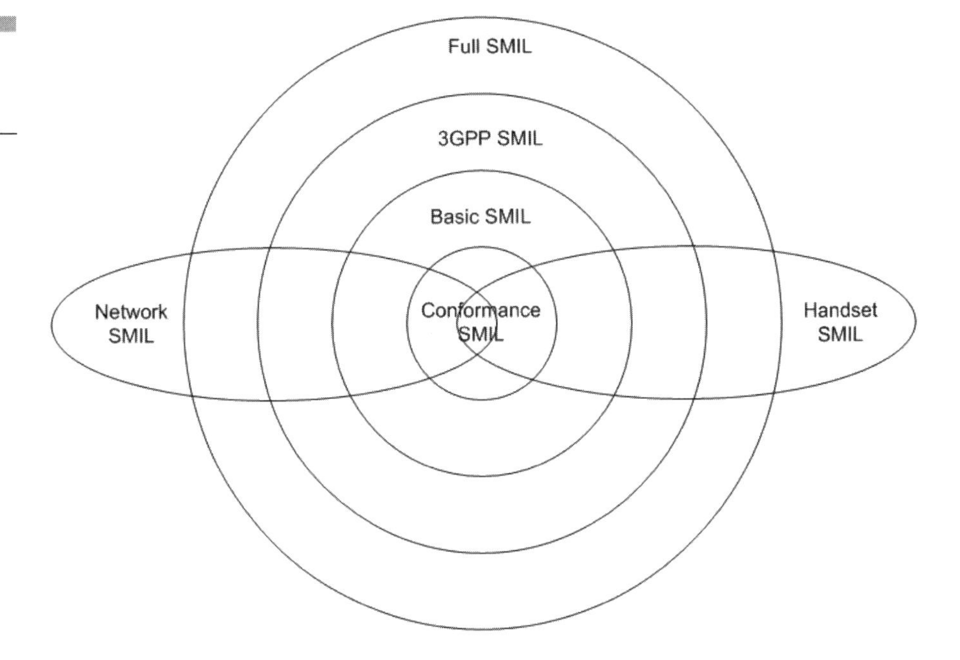

Figure 4.2 Relationship of variants of SMIL.

The layout language for MMS is 3GPP SMIL, as defined in 3GPP TS 26.234. 3GPP SMIL is a proper superset of W3C Basic SMIL 2.0 and a proper subset of W3C Full SMIL 2.0. Conformance SMIL is a proper subset of W3C Basic SMIL 2.0 and, therefore, a proper subset of 3GPP SMIL. Table 4.1 compares the three documented variants of SMIL that you may come across when preparing your MMS message.

TABLE 4.1 Tags in 3GPP, Basic, and Conformance SMIL

TAG	3GPP SMIL	W3C Basic SMIL 2.0	Conformance SMIL
a	X	X	
anchor		X	
animateColor			
animateMotion			
animation	X	X	

(continued on next page)

TABLE 4.1
Tags in 3GPP, Basic, and Conformance SMIL (continued)

TAG	3GPP SMIL	W3C Basic SMIL 2.0	Conformance SMIL
area	X	X	
audio	X	X	X
body	X	X	X
brush			
customAttribute			
customtest			
excl			
head	X	X	X
img	X	X	X
layout	X	X	X
meta	X		X
metadata	X		
mouseFollow			
par	X	X	X
param			
pim			X
prefetch	X		
priorityClass			
rdf			
ref	X	X	X
region	X	X	X
regPoint			
root-layout	X	X	X
seq	X	X	
set			
smil	X	X	X
switch	X	X	

(continued on next page)

TABLE 4.1 Tags in 3GPP, Basic, and Conformance SMIL (continued)

TAG	3GPP SMIL	W3C Basic SMIL 2.0	Conformance SMIL
text	X	X	X
textstream	X	X	
topLayout			
transition	X		
transitionFilter			
video	X	X	

MMS Conformance SMIL

Even though it is not very creative, we will start our consideration of MMS construction with Conformance SMIL. It is the distilled essence of SMIL. As illustrated in Figure 4.3, an MMS Conformance SMIL message consists simply of a series of slides. Each slide can have only two regions. One region can contain a picture, and the other region can contain some text. You can put an audio track under each slide, but an audio track cannot span more than one slide.

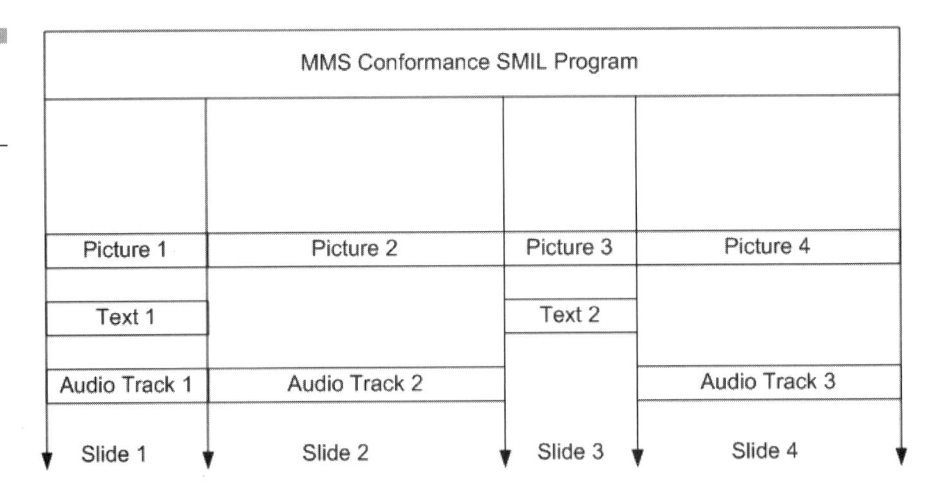

Figure 4.3 Structure of a conformance MMS.

The following is the SMIL script that describes the MMS in Figure 4.3.

```
<smil>
    <head>
        <layout>
            <root-layout width="101" height="80">
            <region id="Image" width="101" height="40" left="0"
             top="0"/>
            <region id="Text" width="101" height="40" left =
             "0" top="40"/>
        </layout>
    </head>
    <body>
        <par dur="10000ms">
            <img src="Picture1.jpg" region="Image" />
            <text src="Text1.txt" region="Text"/>
            <audio src="Audio1.amr" />
        </par>
        <par dur="10000ms">
            <img src="Picture2.jpg" region="Image" />
            <audio src="Audio2.amr" />
        </par>
        <par dur="10000ms">
            <img src="Picture3.jpg" region="Image" />
            <text src=""Text2.txt" region="Text" />
        </par>
        <par dur="10000ms">
            <img src="Picture4.jpg" region="Image" />
            <audio src="Audio2.amr" />
        </par>
    </body>
</smil>
```

The tags of Conformance SMIL, along with the attributes that can be used with them, are given in Table 4.2.

Using the "begin" and "end" attributes, a media element can begin after the display of a slide begins, or end before the display of the slide ends. It is also possible, using multiple regions, that each slide has a different placement of the picture and the text. "alt" is some alternative text that should be displayed if the handset cannot handle the media format of the "src."

Conformance MMS messages come with some other restrictions too. These are listed in Table 4.3. The most onerous restriction is

TABLE 4.2
Tags and Attributes in Conformance SMIL

Tag	Attributes
smil	head, body
meta	name, content
head	layout
body	par
region	left, top, height, fit, id
root-layout	width, height
text	src, region, alt, begin, end
img	src, region, alt, begin, end
audio	src, region, alt, begin, end
ref	src, region, alt, begin, end
par	dur

that a Conformance MMS message cannot be larger than 30 kilobytes—not each slide, but the entire MMS message. In order to calibrate your expectations, the beginners "Hello, world" MMS developed in Chapter 3 came out to be about 20 kilobytes.

TABLE 4.3
Conformance SMIL Restrictions

Aspect	Restriction
Screen size	160 horizontal pixels by 120 vertical pixels
Message size	Less than 30 kilobytes
Image formats	JPEG, GIF87a, GIF89a, WBMP
Audio formats	AMR
Text formats	US-ASCII, UTF-8, UTF-16
PIM	vCalendar, vCard

The rigid structure of Conformance SMIL messages may be satisfactory for greeting cards, advertisements, birth announcements, and party invitations. Network operators hope that this may be all that is needed to jump-start the consumer use of MMS. It is unlikely that we will see any new and innovative "killer" MMS applications rendered within the constraints of Conformance SMIL.

MMS Construction

MMS construction consists of the following three tasks:

1. Compose: Create the SMIL script for the message.
2. Transform: Prepare each of the message's media elements.
3. Encode: Prepare the whole mess(age) for sending.

Due to the unproven business potential of the MMS market, not to mention the small number of MMS application developers, few tools available are specifically built for MMS construction. As a result, you will find yourself using general-purpose tools and adapting them to the MMS construction tasks. For this reason, and because there is not as yet a common agreement on who does what on this construction site, the software tools available for each of the MMS construction tasks do not fit together terribly well.

Compose

Table 4.4 lists a number of different tools for tackling the MMS composition task. The output of this task has to be an SMIL script that can be understood by the handset to which you are sending the MMS. Therefore, you have to worry not only about designing the MMS so that it delights the recipient, but also about keeping all the Multimedia Messaging Service Centers (MMSCs) along the way and the handset happy.

The handset manufacturers' tools are very attractive because they are free and are highly focused on MMS message construction. Unfortunately, the handset manufacturers go out of their way to make sure you cannot use the output from their tools for anything other than viewing on their handset simulators. The Ericsson tool generates MMS messages that can only be played back on special Ericsson handsets and SMIL that is not even close to standard

TABLE 4.4

MMS Message Composers to Create SMIL Scripts

Tool Category	Description	Examples
MMS composers provided by handset vendors	Very few features; targeted toward capabilities of particular handsets; include limits on commercial use; free for downloading.	Ericsson MMS Composer, Nokia Developer's Suite for MMS, Sony Ericsson MMS Home Studio
MMS portals	Very expensive; made to be sold to network operators, not to application developers; assume a direct connect to the MMSC; massive feature set; remote storage of content.	Liquid Air Labs, WapOneLine, mi4e, internexium, mforma, alatto
Multimedia editors	Difficult to stay inside the MMS subset of SMIL; no generation of MIME or MM1 headers or WSP encoding; great interface to media files.	Oratrix GriNS Pro, MacroMedia HomeSite, RealSystem G2
Web authoring tools	Difficult to constrain to the MMS SMIL DTD; hard to synchronize individual images with a continuous time line; overkill for the task.	Adobe GoLive, Macromedia Dreamweaver
WML, i-mode, and cHTML editors	Don't know about SMIL, MIME, or MM1; little or no interface to media files; many are free.	Visual InterWAP, Rasquares Wap Pro 2
Online composers	Hard to capture SMIL; need to upload media elements.	Zidango, Liquid Air Labs, alatto
Text composers	Build the SMIL yourself.	Notepad, emacs

SMIL. The Nokia Developer's Suite for MMS must be used with the Adobe GoLive editor, and generates MMS messages encoded for the Nokia proprietary MMSC interface. Basically, these tools are dead ends from the point of view of a serious MMS application developer.

MMS portal software is also focused on creating MMS messages, but is built and priced to be sold to network operators and existing mega-portals like Yahoo!. The theory is that you will pay per message to develop your MMS application. It is like buying a C compiler and paying for each program you compile. MMS application developers are not considered to be customers by the creators of these composers.

The multimedia editors, particularly those built to edit full bore SMIL, come closest to being industrial-strength MMS development tools. They have a fully developed sense of time synchronization and can readily handle many formats of graphic and audio files (except adaptive multirate [AMR] audio files, which is what MMS likes the best). You cannot constrain these tools to a subset of SMIL, nor can you have the tools create the multipurpose Internet mail extension (MIME) encapsulation, MM1 headers, or wireless session protocol (WSP) encoding that MMS needs.

Some of the major Web authoring tools such as Adobe GoLive and Macromedia Dreamweaver claim to support SMIL and MMS messaging, but as with the multimedia editors, these tools cannot be constrained to the any of the MMS subsets of SMIL. Handsets and MMSCs are a lot less forgiving than desktop Web browsers. Also, like the multimedia editors, these tools do not generate the MIME encapsulations, MM1 headers, or WSP encoding. All they generate is the SMIL program. Web authoring tools also have a very weak sense of time synchronization, if any at all.

Transform

No matter what tool or tools you pick to compose the SMIL of your MMS messages, you will need a tool kit full of media editing and conversion tools to format the media elements that go into your message. The sound bite you want to use will be too long, sampled at too high a rate, and in WAV rather than AMR. The images will need to be clipped, chopped, and converted into JPEG or GIF. The clip art is in BMP rather than PNG. Table 4.5 gives some examples of these tools, along with URLs where you can find more.

In preparing your media for inclusion in an MMS message, the question that arises immediately is, "What is the target?" If you know the detailed technical capabilities of the handset your customer is carrying, then naively it would seem that you should tailor your media for that handset. If you don't know these details or you have no idea what handset your customer is carrying, you are still in the fog.

TABLE 4.5
MMS Media Element Editors

Tool Category	Description	Examples
Graphics editing and transformation tools	Transform pictures from one format to another; change the parameters of a given format; clip, crop, and resize images.	HyperSnap, PaintShop Pro fCoder ImageConverter Plus, Adobe PhotoShop, and lots more at www.tucows.com and graphicssoft.about.com
Audio editing and transformation tools	Transform audio tracks from one format to another; change the amplitude, sampling rate, and other parameters of the audio track; clip and crop the audio track.	Gold Wave, SPOTxdePro, Cool Edit Pro, and much more atwww.tucows.com and www.sonicspot.com

Network operators do have the wherewithal to determine what handset the recipient is carrying, and current thinking is that they will perform transcoding and content adaption services that will transform your MMS message into a message that is optimal for the receiving handset. A long discussion of the pros and cons of such a transformation service is at the end of Chapter 3, but for the moment, let's assume that such a transcoding service is in use. How does this help you solve your problem of preparing your media for inclusion in the MMS message?

In fact, it makes your problem worse, not better. Previously, if you knew the technical details of your customer's handset, you could prepare your message for optimal playback on the handset. Now, an unknown and unknowable transformation will be applied to your message. It is possible that the final result would be better if you didn't tune the message for the handset, but rather gave it to the network in a form that was easily handled by the transcoding engine. This is certainly what you would want to do if you did not know the handset at all.

But what is the generic format for each media element? Is a square image a better bet than a landscape or portrait sizing? Should you always use as many colors as possible, or as few? What color transformations will be used? Will the red that you send be the red that is received? And what will happen to the audio tracks?

Should you clip the highs and lows before you send, or is it better to include them?

These questions and many more like them are critically important to multimedia application developers and content providers. Disney will not take kindly to having Mickey's ears turned from black to gray, or having Goofy come out sounding like Minnie. These are not, however, as yet questions to which either the network operators or the MMS equipment providers have shown any sensitivity. Preparing your media elements for inclusion in an MMS message today is a shot in the dark, and there is every indication that the telecommunications industry wants to keep it that way.

Encode

After you have finished the first two MMS construction steps, you have an SMIL script that says how to play back the media elements, and you have the media elements themselves. You now have to package this all up and send it off to the network.

How you encode the MMS message for delivery to the network depends on to whom you plan to hand it off. The standards describe two encodings: an ASCII format and a binary format. The ASCII format is a MIME encoding of the message, and the binary format is an MMS encapsulation protocol encoding based on WSP of the MIME encoded message. Both of these formats are typically handed off with hypertext transfer protocol (HTTP) or simple object access protocol (SOAP) headers that provide information about who gets the message, who it is from, and what the subject is. The details of these headers are given in later sections, where we describe particular interfaces.

If MMS gets any traction, services will pop up that accept MMS messages in other formats. It is easy to imagine, for example, a service that accepts an e-mail that contains the SMIL script and the media elements as attachments and does the MIME or WSP encoding and forwarding for you. Or, imagine a service that accepts a ZIP file containing the SMIL script and the media elements and does the same thing. Table 4.6 lists encoding tools.

TABLE 4.6 MMS Encoding Tools

Tool Category	Description	Examples
MIME encoding	Take an SMIL script and the media elements it references and create the MIME encoding of the MMS message ready for sending to an MMSC.	MIME encoding subroutine libraries in Perl, tcl, Python, Visual Basic, Java, C, etc. Hunney Software MIME++
WSP encoding	Take an SMIL script and the media elements it references and create the WSP encoding of the MMS message ready for sending to a WAP gateway.	mmscomp from NowSMS, Nokia MMS Java Library
MMS platforms and libraries		Nowsms.com, M-enable from brainstorm

SMIL Players and MMS Handset Simulators

After you compose, assemble, and encode your MMS message, it is natural to want to see how it will appear to your recipient. This can be accomplished by either using a handset simulator or SMIL playback program in your Windows development environment or by getting your MMS onto a real MMS handset. In the authors' experience, the latter approach—actually getting the MMS onto one or more MMS handsets—is far superior to trying to work with handset simulators or SMIL players.

There are two primary reasons why testing with a real MMS handset beats using a handset simulator. First, you want to tune the pictures and sound in the MMS to the capabilities of the MMS handset, not the capabilities of your desktop computer. MMS handset simulators make no attempt to simulate the playback characteristics of the handset hardware. They really just play back your MMS message in a handset skin.

Second, the fact that an SMIL player or handset simulator accepts your MMS message is no guarantee that your MMS will be accepted by a real MMS handset. In fact, there are SMIL players and handset simulators that don't accept MMS encodings that would be accepted by a handset, and only accept encodings that won't be accepted by a handset. In other words, the players and emulators are leading you down a garden path. Nevertheless, for those who like to work on their desktop machines for as long as possible, Table 4.7 lists some SMIL players and MMS handset simulators.

TABLE 4.7
SMIL Players and MMS Handset Simulators

Name	Source
Nokia handset and MMSC emulators	www.forum.nokia.com
SMIL Player from InterObject	www.inobject.com
GRiNS Player from Oratrix	www.oratrix.com
Real Networks Real One Player	www.realnetworks.com
Internet Explorer 6.0	www.microsoft.com

It would be optimal to hitch up your MMS handset to a communication port on your workstation and simply send it MMS messages to see how they looked. In fact, special versions of the Sony Ericsson T68i support this, but they are very hard to find.

The alternative is simply to send yourself MMS test messages, but this can become very expensive. Some networks and some older MMS handsets allow you to get an MMS onto your phone from a wireless application protocol (WAP) page, but since this circumvents the MMS architecture, newer handsets block this approach.

Another approach is to join a developer program and use the facilities of the program to test your MMS application. The major developers' programs are summarized below.

Application Developer Programs

Everyone in the MMS "food chain" has realized that without content and without applications, MMS will fail. While many hopeful analogies are made to short messaging service (SMS), the two services are fundamentally different. SMS succeeded because it was person-to-person and it was cheaper than making a voice call. Neither of these is a property of MMS.

Content is messy. Everyone wants to make money passing it along, but nobody wants to actually create it. At the end of the day, you, the MMS application developer and MMS content provider, will determine the success or failure of MMS.

With this realization in mind, handset manufacturers, infrastructure providers, and even some network operators sponsor application developer programs. The idea is to make MMS technology more approachable and to make it easier to create MMS applications. What is not covered in developer programs is getting your application to market.

Almost all programs provide introductory "How to Get Going" guides, and most provide easy access to the various standards and specifications that underlay MMS. Beyond these basics, the programs vary widely. Table 4.8 lists the properties of some of the major MMS application developers' programs, as of early 2003. These programs are constantly changing and improving, however, so it is best to visit the Web site and find out what is available.

TABLE 4.8 Application Developers' Programs

Sponsor	URL	Forums	Software Tools	Product Discounts	Server Access	Application Distribution
Handset Manufacturers						
Nokia	www.forum.nokia.com	✔	✔		✔	
Ericsson	www.ericsson.com/mobilityworld	✔	✔			
Qualcomm	www.qualcomm.com/brew	✔	✔	✔	✔	✔
Infrastructure Providers						
Comverse	developer.comverse.com		✔			
Liquid Air Labs	www.liquidairlab.de		✔		✔	
Logica	www.logica.com		✔			
Internexium	www.internexium.com		✔		✔	

(continued on next page)

TABLE 4.8 Application Developers' Programs (continued)

Sponsor	URL	Forums	Software Tools	Product Discounts	Server Access	Application Distribution
Mforma	devnet.mforma.com:8080					✔
Openwave	developer.openwave.com	✔	✔		✔	
Network Operators						
AT&T Wireless Data	www.attws.com/developer	✔			✔	
Cingular	alliance.cingularinteractive.com	✔	✔	✔	✔	
SprintPCS	developer.sprintpcs.com	✔				✔

Other Useful Software Tools

A key component of almost all the developers' programs is the availability of software tools that are free for downloading. Some of these tools are little more than executable demos, but others are good in their own right and can serve as starting places for your own applications. The Nokia NMSS SDK API and library, for example, is a great way to learn about the details of MMS encoding.

In addition to the software available through the various developers' programs, a growing amount of software is being created by other MMS developers and made available on the Web. Table 4.9 lists some of this software, along with a description and the URL where you can get them.

TABLE 4.9 Useful Software Tools for MMS Application Development

Tool	URL	Description
SMS/MMS/WAP Push Gateway	www.nowsms.com	Fully functional MMSC with a very liberal free trial program. Great DOS utility for building MMS from the SMIL script.
GSM::SMS	www.tektonica.com/projects/gsmsms	An excellent Perl module for sending all sorts of SMS messages, including MMS notifications.

(continued on next page)

TABLE 4.9 Useful Software Tools for MMS Application Development (continued)

Tool	URL	Description
Kannel	www.kannel.3glab.org	Open source SMS and WAP gateway that does MMS notifications.
Nokia MMS Java Library	www.developers.forum.com	NMSS SDK for WSP encoding MMS messages; such as m-retreve.conf. Doesn't do the MMS notification message.
mmsCheck	www.nccglobal.com	Reviews your MMS message and specifies on which handsets it will play back successfully.

Summary

The bottom line is that you should plan to assemble your own tool set for building and testing your MMS applications. There are many useful bits and pieces out there, but you have to make them work together to produce MMS messages that do what you want them to do.

If MMS takes off, then perhaps there will be some MMS composers that application developers can use out of the box. Right now, the composers that are available are either so vendor-specific and crippled that the output is useless or so simplistic that you cannot build anything more complex than a birthday card, or both. So be prepared to assemble your own MMS composition toolbox.

CHAPTER 5

Media Formats

Some of the hardest decisions multimedia messaging service (MMS) application builders have to make concern the media formats that will be used in their MMS messages. Should the pictures be rendered in SVG, GIF, or JPEG? Will the audio come out best in iMelody, MIDI, or AMR?

You, the application designer and builder, have to take into account not only the format of the original source of the content, the size of the resulting MMS message, and the capabilities of the phone on the far end that will be rendering your message, but also what various network operators may do to your message along the way.

The definitive documents for MMS media formats are 3GPP TS 26.140, "Multimedia Messaging Service (MMS): Media formats and codecs" and 3GPP TS 26.234, "Transparent end-to-end packet switched streaming service (PPS): Protocols and codecs." A format is the representation of the media content as a sequence or stream of bits and bytes. A codec is the hardware and software combination that converts the bits and bytes back into an analog format that your body can consume.

Table 5.1 is the list of media formats that have been discussed in one forum or another as possibilities for inclusion in MMS messages.

TABLE 5.1 MMS Media Formats

Media	File Extension	MIME Content Type
Text		
IANA MIBenum 3	`txt`	`plain/text; charset="us-ascii"`
IANA MIBenum 100	`txt`	`plain/text; charset="utf-8"`
IANA MIBenum 1000	`txt`	`plain/text; charset="utf-16"` or `plain/text; charset="ISO-10646-ucs-2"`
IANA MIBenum 1015	`txt`	`plain/text; charset="utf-16"` or `plain/text; charset="iso-10646-ucs-2"`
IANA MIBenum 4 (ISO 8869-1)	`txt`	`plain/text; charset="iso-8869-1"`
ISO 8859-1	`txt`	`plain/text; charset="iso-8859-1"`

(continued on next page)

TABLE 5.1
MMS Media Formats (continued)

Media	File Extension	MIME Content Type
Graphics		
GIF87	gif	image/gif
GIF89a	gif	image/gif
JPEG	jpg	image/jpg
WBMP	wbmp	image/vnd.wap.wbmp
JPEG2000	jpg	image/jpg
PNG	png	image/png
SVG	svg, svgz	image/svg
OTA BMP (Nokia Smart Messaging)	ota	image/x-ota-bitmap
BMP	bmp	image/x-bmp
MBM	mbm	image/x-epoc.mbm
TIFF-F	tif, tiff	image/tiff
Audio		
Ringtone		application/vnd.nokia.ringing-tone
Melody	mel	text/x-vmel
eMelody	emy	text/x-emelody
iMelody	imy	text/x-imelody
AMR-Narrow Band	amr	audio/amr
AMR-Wide Band	amr	audio/amr
EVRC	evr	audio/evrc
SMV	Smv	audio/smv
MIDI	mid, midi	audio/x-midi
SP-MIDI	mid, midi	audio/sp-midi
SMAF	mmf	application/x-smaf
WAV	wav	audio/x-wav

(continued on next page)

TABLE 5.1 MMS Media Formats (continued)

Media	File Extension	MIME Content Type
MPS	mps	audio/mps
PSION	wve	audio/x-sibo, audio/x-sibo-wve
RMF	rmf	audio/rmf, audio/x-rmf, audio/x-beatnic-rmf
RealAudio	ra, ram, rm, rmp	audio/x-pn-realaudio
Sun/Next	au	audio/basic
Windows Media Audio	wma	audio/x-ms-wma
QCELP	qcp	audio/qcp
Video		
AVI	avi	video/avi
3GPP	*3gp*	*video/3gp*
MPEG-4	mpeg	video/mpeg
ITU-T H.263	h263	video/x-h263
QuickTime	qt	video/quicktime
Miscellaneous		
vCard	vcd	text/x-vcard, text/directory; profile="vcard"; charset=<whatever it is>
vCalendar	vcl	text/x-vCalendar, text/calendar, text/directory;profile="vcalendar"; charset=whatever it is>
Java Script	jar jad	application/java-archive text/vnd.sun.j2me.app-descriptor

Media formats can generally be classified as low resolution or high resolution. This classification for the most part also distinguishes between synthetic and real. Low-resolution formats transmit quickly and cheaply, but only render a low-dimensional sketch of the object. High-resolution formats transmit many times more bits than low-resolution formats, but cover the full range and complexity of the sensory spectrum. Table 5.2 gives classifications of some popular media formats.

TABLE 5.2 Low- versus High-Resolution Media Formats

	Low Resolution Synthetic Projection	High-Resolution Realistic Full-Dimension
Text	ASCII	Unicode
Audio	iMelody, MIDI	AMR, QCELP, SMV, WAV, MPEG
Graphics	SVG	GIF, JPEG
Video	AVI	H.263, MPEG4

On some low-end MMS handsets, it is not only a waste of time and expense to send a high-resolution format, but the result can actually be less acceptable than if a low-resolution format had been used. Table 5.3 is a list of the formats accepted by some popular MMS handsets.

TABLE 5.3 Properties of Some Early MMS Handsets

Handset	Supported Media Formats	Colors	Screen Size (h × v pixels)	Maximum MMS Size	Memory Size	Comment
Sony Ericsson t68i	GIF87, GIF89a, JPG, JPEG, WBMP, EMY, IMY, MEL, AMR, VCS, VCF, VNT, THM, URL	256	101 × 80	30 kB	1 MB	A real workhorse, but limited color palette.
Nokia 7650	BMP, GIF87, GIF89a, JPEG, MBM, PNG, WBMP, TIFF-F, AMR, MIDI	4,096	176 × 208	100 kB	3.6 MB	Extremely limited SMIL support; can support video/3GGP with an add-on application.
Motorola T720i	PNG, WBMP, GIF, MIDI, IMY, JPEG	4,096	120 × 160	30 kB	10 MB	Extremely limited SMIL support.
Panasonic GD87	PNG, JPEG, GIF89a, BMP, WBMP, MIDI, AMR	65,536	132 × 176	50 kB	780 kB	Truly beautiful screen; early versions have bugs.

To make sure your MMS gets through, the theory is to confine yourself to the formats listed in the MMS Conformance Document, Version 2.0.0. These are listed in Table 5.4. Be warned, however, that there are handsets that claim conformance to the Conformance Document but do not conform to it.

TABLE 5.4
MMS Conformance Document V2.0.0 Media Formats

Text
IANA MIB enum3 (us-ascii)
IANA MIB enum 106 (utf-8 Unicode)
IANA MIB enum 1000 (utf-16 Unicode)
ISO 8859-1 (Latin 1 Glyphs)
Graphics
GIF87
GIF89a
WBMP
Audio
AMR
Miscellaneous
vCalendar version 1.0
vCard version 2.1

The biggest drawback to the MMS Conformance Document list is that adaptive multirate (AMR) is the only audio format supported. Not only is it difficult to translate non-AMR audio gracefully into AMR, but if you only want to play some synthetic notes or sounds such as those available in iMelody or MIDI, you get a file that is much larger than you need. AMR is also not suitable for use as a ring tone in most devices, where you would use the Nokia RTTTL format, iMelody, MIDI, or SMAF. Based on the success of Short Messaging Service (SMS) delivery of ring tones, MMS would make an ideal delivery platform for more sophisticated ring tones; however, for the foreseeable future, different ring tone formats will be preferred by different handset vendors.

As you would expect, the first preference in handsets when it comes to audio is to use the codec that is already in the handset for speech. Some of these codecs are listed in Table 5.5. All of these codecs are very, very highly optimized for human speech, simply because reduced bit rate translates immediately into the ability to handle more calls. It is to be seen if the need to handle multimedia audio, other than human speech, will carry enough weight to start to loosen up these optimizations.

TABLE 5.5
Audio Codecs in Various Mobile Telephone Technologies

Codec	System
VSELP	TDMA
ACELP	TDMA
GSM	GSM
AMR	3GPP
CELP	CDMA
QCELP-13	CDMA
EVRC	3GPP2
SVC	IMT-2000

One scenario for the future is that mobile telephones converge on a common audio codec. Scalable vector graphics (SVG) is certainly a candidate. The authors are not aware of any tests focused on SVG's ability to carry a Bach fugue or Nusrat Fateh Ali Khan's latest hit.

Other than the native voice codec, the 3GPP and 3GPP2 vision of MMS, including architecture and media formats, is about the same. Indeed, it is so much the same that interoperability between the two networks in the matter of MMS messaging is not only considered, but almost assumed. This does not mean that there still will not be a few interoperability meetings. 3GPP2 is considering the use of data calling tone (DCT) as a still image format, and could go with the mobile profile of extensible hypertext markup language (XHTML) rather than synchronized multimedia integration language (SMIL) as the scripting language, but these differences, as with the vocoder issue, can be smoothed over with transcoding.

Determining What a Handset Handles

It can be frustrating to find out exactly what media formats and what SMIL elements and attributes a particular MMS handset will handle. Some handset manufacturers have published standardized user agent profile (UAProf) descriptions of their handsets. Appendix C is the schema for one of the many user agent profiles for MMS mobile phones and the following listing is the UAProf for the Panasonic GD87. A very useful collection of the currently available mobile phone UAProfs can be found at http://w3development.de/rdf/uaprof_repository.

The primary purpose of UAProf and the World Wide Web Consortium's (W3C's) composite capability/preference profiles (CC/PP) is to provide servers with a standardized and machine readable way of answering the question, "What can this thing that is hitting on me do?" but they are just as good at answering this question for application developers.

```
<?xml version="1.0"?>
<!-- ========================================================= -->
<!--                                                            -->
<!-- Copyright (c) 2002                                         -->
<!-- Matsushita Communication Industrial Co., Ltd.              -->
<!--                                                            -->
<!-- ========================================================= -->
<!--                                                            -->
<!-- File Name:                                                 -->
<!--  GD87R1.xml                                                -->
<!-- Purpose:                                                   -->
<!--  This file contains the User Agent Profile for the GD87    -->
<!--  Currently at version R1.                                  -->
<!-- History:                                                   -->
<!--  25thOct02 change the schema from "ccppschema-20000405"    -->
<!--              to "ccppschema-20020710"                      -->
<!--     28thOct02 modify the URL of MMS schema                 -->
<!--                                                            -->
<!-- ========================================================= -->
```

```
<rdf:RDF xmlns:rdf="http://www.w3.org/1999/02/22-rdf-syntax-ns#"
xmlns:rdfs="http://www.w3.org/2000/01/rdf-schema#"
xmlns:prf="http://www.wapforum.org/profiles/UAPROF/
      ccppschema-20020710#">
     <rdf:Description rdf:ID="UAPROF">
<!--  **************************************************   -->
<!--  ***** Component: HardwarePlatform *****   -->
          <prf:component>
               <rdf:Description rdf:ID="HardwarePlatform">
               <rdf:type rdf:resource="http://www.wapforum.org/
                    profiles/UAPROF/ccppschema-20020710#
                    HardwarePlatform"/>
                    <prf:BitsPerPixel>16</prf:BitsPerPixel>
                    <prf:ColorCapable>Yes</prf:ColorCapable>
                    <prf:ImageCapable>Yes</prf:ImageCapable>
                    <prf:InputCharSet>
                         <rdf:Bag>
                              <rdf:li>ISO-8859-1</rdf:li>
                              <rdf:li>UTF-8</rdf:li>
                         </rdf:Bag>
                    </prf:InputCharSet>
                    <prf:Keyboard>PhoneKeypad</prf:Keyboard>
                    <prf:Model>GD87</prf:Model>
                    <prf:NumberOfSoftKeys>3</prf:
                         NumberOfSoftKeys>
                    <prf:OutputCharSet>
                         <rdf:Bag>
                              <rdf:li>ISO-8859-1</rdf:li>
                              <rdf:li>UTF-8</rdf:li>
                         </rdf:Bag>
                    </prf:OutputCharSet>
                    <prf:PixelAspectRatio>1x1</prf:
                     PixelAspectRatio>
                    <prf:ScreenSize>132x176</prf:ScreenSize>
                    <prf:ScreenSizeChar>16x8</prf:ScreenSizeChar>
                    <prf:StandardFontProportional>Yes</
                         prf:StandardFontProportional>
                    <prf:SoundOutputCapable>Yes</
                         prf:SoundOutputCapable>
                    <prf:TextInputCapable>Yes</
                         prf:TextInputCapable>
```

```
                <prf:Vendor>Panasonic</prf:Vendor>
            </rdf:Description>
        </prf:component>
<!--  ************************************************************   -->
<!--  ***** Component: SoftwarePlatform *****   -->
        <prf:component>
            <rdf:Description rdf:ID="SoftwarePlatform">
            <rdf:type rdf:resource="http://www.wapforum.org/
             profiles/UAPROF/ccppschema-
             20020710#SoftwarePlatform"/>
                <prf:cceptDownloadableSoftware>No</
                    prf:AcceptDownloadableSoftware>
                <prf:CcppAccept>
                    <rdf:Bag>
                        <rdf:li>text/plain</rdf:li>
                        <rdf:li>text/css</rdf:li>
                        <rdf:li>text/x-server-parsed-
                                html</rdf:li>
                        <rdf:li>text/vnd.wap.wml</rdf:li>
                        <rdf:li>text/vnd.wap.wmlscript
                                </rdf:li>
                        <rdf:li>image/gif</rdf:li>
                        <rdf:li>image/png</rdf:li>
                        <rdf:li>image/jpeg</rdf:li>
                        <rdf:li>image/vnd.wap.wbmp</rdf:li>
                        <rdf:li>image/bmp</rdf:li>
                        <rdf:li>audio/midi</rdf:li>
                        <rdf:li>audio/mid</rdf:li>
                        <rdf:li>audio/imelody</rdf:li>
                        <rdf:li>audio/amr</rdf:li>
                        <rdf:li>audio/sp-midi</rdf:li>
                        <rdf:li>application/x-pmd</rdf:li>
                        <rdf:li>application/vnd.wap.wmlc</
                                rdf:li>
                        <rdf:li>application/vnd.wap.wbxml</
                                rdf:li>
                        <rdf:li>application/vnd.wap.
                                wmlscriptc</rdf:li>
                        <rdf:li>application/vnd.wap.
                                multipart.mixed</rdf:li>
                        <rdf:li>application/vnd.wap.
```

```
                                        wtls-ca-certificate</
                                        rdf:li>
                                <rdf:li>application/wml+xml</
                                        rdf:li>
                                <rdf:li>application/xhtml+xml</
                                        rdf:li>
                            </rdf:Bag>
                        </prf:CcppAccept>
                        <prf:CcppAccept-Charset>
                            <rdf:Bag>
                                <rdf:li>ISO-8859-1</rdf:li>
                                <rdf:li>UTF-8</rdf:li>
                                <rdf:li>US-ASCII</rdf:li>
                            </rdf:Bag>
                        </prf:CcppAccept-Charset>
                        <prf:CcppAccept-Encoding>
                            <rdf:Bag>
                            <rdf:li>base64</rdf:li>
                            </rdf:Bag>
                        </prf:CcppAccept-Encoding>
                    </rdf:Description>
                </prf:component>
<!--    *****************************************************    -->
<!--    ***** Component: NetworkCharacteristics *****    -->
                <prf:component>
                    <rdf:Description rdf:ID="NetworkCharacteristics">
                    <rdf:type rdf:resource="http://www.wapforum.org/
                        profiles/UAPROF/ccppschema-
                        20020710#NetworkCharacteristics"/>
                        <prf:SecuritySupport>
                            <rdf:Bag>
                                <rdf:li>WTLS-1</rdf:li>
                                <rdf:li>WTLS-2</rdf:li>
                            </rdf:Bag>
                        </prf:SecuritySupport>
                        <prf:SupportedBearers>
                            <rdf:Bag>
                                <rdf:li>OneWaySMS</rdf:li>
                                <rdf:li>CSD</rdf:li>
                                <rdf:li>GPRS</rdf:li>
                            </rdf:Bag>
```

```
                </prf:SupportedBearers>
            </rdf:Description>
        </prf:component>
<!--  ****************************************************   -->
<!--  ***** Component: BrowserUA *****   -->
        <prf:component>
            <rdf:Description rdf:ID="BrowserUA">
            <rdf:type rdf:resource="http://www.wapforum.org/
                profiles/UAPROF/ccppschema-
                20020710#BrowserUA"/>
                <prf:BrowserName>Panasonic</prf:BrowserName>
                <prf:FramesCapable>No</prf:FramesCapable>
                <prf:TablesCapable>Yes</prf:TablesCapable>
            </rdf:Description>
        </prf:component>
<!--  ****************************************************   -->
<!--  ***** Component: WapCharacteristics *****   -->
        <prf:component>
            <rdf:Description rdf:ID="WapCharacteristics">
            <rdf:type rdf:resource="http://www.wapforum.org/
                profiles/UAPROF/ccppschema-
                20020710#WapCharacteristics"/>
                <prf:WapDeviceClass>C</prf:WapDeviceClass>
                <prf:WapVersion>1.2.1</prf:WapVersion>
                <prf:WmlDeckSize>12000</prf:WmlDeckSize>
                <prf:WmlScriptLibraries>
                    <rdf:Bag>
                        <rdf:li>Lang</rdf:li>
                        <rdf:li>Float</rdf:li>
                        <rdf:li>String</rdf:li>
                        <rdf:li>URL</rdf:li>
                        <rdf:li>WMLBrowser</rdf:li>
                        <rdf:li>Dialogs</rdf:li>
                    </rdf:Bag>
                </prf:WmlScriptLibraries>
                <prf:WmlScriptVersion>1.2</
                    prf:WmlScriptVersion>
                <prf:WmlVersion>1.3</prf:WmlVersion>
                <prf:WtaiLibraries>
                <rdf:Bag>
                <rdf:li>WTA.Public.makeCall</rdf:li>
```

```
                <rdf:li>WTA.Public.sendDTMF</rdf:li>
                <rdf:li>WTA.Public.addPBEntry</rdf:li>
              </rdf:Bag>
            </prf:WtaiLibraries>
            </rdf:Description>
        </prf:component>
<!--  *****************************************************   -->
<!--  ***** Component: PushCharacteristics *****   -->
        <prf:component>
            <rdf:Description rdf:ID="PushCharacteristics">
            <rdf:type rdf:resource="http://www.wapforum.org/
                profiles/UAPROF/ccppschema-
                20020710#PushCharacteristics"/>
                <prf:Push-MsgSize>12000</prf:Push-MsgSize>
            </rdf:Description>
        </prf:component>
<!--  *****************************************************   -->
<!--  ***** Component: MmsCharacteristics *****   -->
        <prf:component>
            <rdf:Description rdf:ID="MmsCharacteristics">
            <rdf:type rdf:resource="http://www.wapforum.org/
                profiles/MMS/ccppschema-
                20010111#MmsCharacteristics"/>
                <prf:MmsMaxMessageSize>50000</
                    prf:MmsMaxMessageSize>
                <prf:MmsMaxImageResolution>132x134</
                    prf:MmsMaxImageResolution>
                <prf:MmsCcppAccept>
                    <rdf:Bag>
                        <rdf:li>text/plain</rdf:li>
                        <rdf:li>image/gif</rdf:li>
                        <rdf:li>image/png</rdf:li>
                        <rdf:li>image/jpeg</rdf:li>
                        <rdf:li>image/vnd.wap.wbmp</rdf:li>
                        <rdf:li>image/bmp</rdf:li>
                        <rdf:li>audio/midi</rdf:li>
                        <rdf:li>audio/mid</rdf:li>
                        <rdf:li>audio/imelody</rdf:li>
                        <rdf:li>audio/amr</rdf:li>
                        <rdf:li>application/smil</rdf:li>
                        <rdf:li>multipart/mixed</rdf:li>
```

```
                    <rdf:li>application/vnd.wap.
                            multipart.related</rdf:li>
                    <rdf:li>application/vnd.wap.
                            multipart.mixed</rdf:li>
               </rdf:Bag>
          </prf:MmsCcppAccept>
          <prf:MmsCcppAcceptCharSet>
               <rdf:Bag>
                    <rdf:li>ISO-8859-1</rdf:li>
                    <rdf:li>ISO-8859-2</rdf:li>
                    <rdf:li>US-ASCII</rdf:li>
                    <rdf:li>UTF-8</rdf:li>
                    <rdf:li>UCS-2</rdf:li>
               </rdf:Bag>
          </prf:MmsCcppAcceptCharSet>
          <prf:MmsVersion>1.0</prf:MmsVersion>
     </rdf:Description>
   </prf:component>
 </rdf:Description>
</rdf:RDF>
```

There is a nice little open source application program interface (API) called delivery context library (DELI)that provides access to CC/PP and UAProf descriptions. You can download it from http://sourceforge.net/projects/delicon/.

UAProfs do not cover elements and attributes the SMIL interpreter in the handset supports, together with any constraints on those elements and attributes. About the only way of doing this is to send a SMIL script to the handset and see what happens. You want to do this anyway, to not only make sure the handset will handle your script, but to see how well it does so.

That a handset supports a media format only really means that it does not roll over and die if you send it the format. It does not say a thing about how well it supports it, either in terms of the quality of the playback or the speed of the playback. The authors have set up an SMS responder that contains a selection of MMS test messages. Table 5.6 lists the MMS test messages that are available as of early 2003. Go to http://www.mobile-mind.com/mmstests.html to get a list of the current MMS test messages and instructions on how to have the test MMS messages sent to your handset.

TABLE 5.6
MMS Test Messages from SMS Responder

Test Identifier	Subject of MMS	MMS Size (bytes)	Comment
	Text Tests		
ascii	ASCII text	602	ASCII text
utf8bom	utf-8 BOM	620	ASCII text as utf-8
utf16le	utf-16 little endian	814	ASCII text as utf-16
utf16be	utf-16 big endian	817	ASCII text as utf-16
names	utf-8 names	3701	Names with utf-8 letters
	Image Tests		
gifdr	Black-and-white GIF89a drawing	5978	Letter G
gifln	Black-and-white GIF89a photo	8426	Lenna
gifba	Colored GIF89a image	8112	Baboon
gif89a	Animated GIF89a image	16241	Dallas
gif16	GIF89a 16-gray bar	1488	16 gray shades
gif256	GIF89a 256-gray bar	2422	80 gray shades
gifblu	GIF89a pure blue	1348	Color purity test
gifgrn	GIF89a pure green	1349	Color purity test
gifred	GIF89a pure red	1347	Color purity test
gifbow	GIF89a rainbow	4263	252 unique colors
jpgdr	Black-and-white JPEG drawing	8641	Letter G
jpgln	Black-and-white JPEG photo	5750	Lenna
jpgba	Colored JPEG photo	8314	Baboon
jpg16	JPEG 16-gray bar	3237	46 gray shades
jpg256	JPEG 256-gray bar	2527	101 gray shades
jpgblu	JPEG pure blue	1156	Color purity test
jpggrn	JPEG pure green	1156	Color purity test
jpgred	JPEG pure red	1154	Color purity test
jpgbow	JPEG rainbow	4518	1,530 unique colors
pngln	Black-and-white PNG photo	5108	Lenna

(continued on next page)

TABLE 5.6 MMS Test Messages from SMS Responder (continued)

Test Identifier	Subject of MMS	MMS Size (bytes)	Comment
pngba	Color PNG photo	15297	Baboon
svgbb	SVG animation	4831	Blue bouncing ball
	Audio Tests		
amr	AMR audio	1751	
mel	Melody audio	550	
emel	eMelody audio	620	
imel	iMelody audio	785	
midi	MIDI audio	3708	
	Miscellaneous Tests		
vcard	vCard file	802	
vcal	vCalendar file	946	

Text

You can be absolutely certain that ASCII (also known as IANA MIBenum 3, ISO 646-US, ANSI_X3.4-1968, ISO_646.irv:1991, and IBM 367) will work. While text-only MMSs might seem like an expensive and high-tech simulation of SMS, a text-only MMS offers at least two useful advantages over SMS: long messages and interactivity.

Practically speaking, the upper limit of an SMS is about 800 characters. This is possible if you use concatenated SMSs and your operator allows you to chain together up to five of them. The typical SMS is 150 characters or less. The MMS Conformance Document states that MMSs up to 30K will be handled by all conforming MMS handsets and MMS infrastructure elements. 30K is also about the size the operators are referring to in MMS introductions. Even taking into account the SMIL and the multipurpose Internet mail extension (MIME) encoding, you can get a lot more text into an MMS than you can into an SMS.

Text, when used with the <a> ("anchor") or <area> linking elements of Basic SMIL 2.0, can provide an easy way to add pick lists and choice menus to MMS messages. A text-only MMS with anchors is similar to an interactive SMS. Rather than recipients having to reply to the SMS and type in their choice, they can simply scroll down on the text MMS and pick. The only problem is that most of the current generation of handsets does not support linking. Nevertheless, since interaction is such a compelling feature of MMS and one that is easily implemented, given the existence of the wireless application protocol (WAP) browser in the handset, one can safely assume that interactive MMSs, including interactive text-only MMSs, are just over the horizon.

There are two ways to achieve MMS menus within today's 3GPP SMIL profile: the anchor or the area. Since neither is implemented in today's handsets, it is not clear which one will be favored or even supported.

In the anchor approach, one makes a different text area on the screen for each menu choice. The following is a SMIL script that illustrates this approach.

```
<smil>
  <head>
    <layout>
      <root-layout width="320" height="240" title="Pick Quick!"/>
      <region id="text1" top="000" left="0" height="60"
       width="320"/>
      <region id="text2" top="060" left="0" height="60"
       width="320"/>
      <region id="text3" top="120" left="0" height="60"
       width="320"/>
      <region id="text4" top="180" left="0" height="60"
       width="320"/>
  </layout>
  </head>
  <body>
    <par dur="50000ms">
        <a href="http://mms.mobile-mind.com/examples/menu1.pl?1">
            <text src="choice1.txt" region="text1"/>
        </a>
        <a href="http://mms.mobile-mind.com/examples/menu1.pl?2">
            <text src="choice2.txt" region="text2"/>
        </a>
        <a href="http://mms.mobile-mind.com/examples/menu1.pl?3">
```

```
          <text src="choice3.txt" region="text3"/>
        </a>
        <a href="http://mms.mobile-mind.com/examples/menu1.pl?4">
          <text src="choice4.txt" region="text4"/>
        </a>
    </par>
  </body>
</smil>
```

We put a different text file into each region to identify the choice. When a pick is made, a script on the sending server is hit with the index of the choice.

An alternative is to use the <area> tag. The example Basic SMIL 2.0 script for a text MMS menu presentation is given below.

```
<smil>
  <head>
    <layout>
      <root-layout width="320" height="240" title="Pick Quick!"/>
      <region id="text" top="0" left="0" height="100%"
       width="100%"/>
  </layout>
  </head>
  <body>
    <par dur="5000ms">
      <area shape="rect" coords="0,0,320,60"
              href="http://mms.mobile-mind.com/examples/
                    menu1.pl?1">
      <area shape="rect" coords="0,60,320,120"
              href="http://mms.mobile-mind.com/examples/
                    menu1.pl?1">
      <area shape="rect" coords="0,120,320,180"
              href="http://mms.mobile-mind.com/examples/
                    menu1.pl?1">
      <area shape="rect" coords="0,180,320,240"
              href="http://mms.mobile-mind.com/examples/
                    menu1.pl?1">
      <text src="choices.txt" region="text" />
    </par>
  </body>
</smil>
```

The coordinates in the area tag are given by:

```
coords-value ::= left-x "," top-y "," right-x "," bottom-y
```

Area linking requires putting each area over the right line in the text file, and this will likely be more unstable over handsets than the anchor approach. Areas are (or more properly, will be) great for making pictures into picks, however, as long as any content adaptation services in the operator network do not get in the way.

ISO 8859-1 is the set of special characters for most of the alphabets in the world, save the Arabic and Asian languages. Figure 5.1 shows this character set. It is part of the MMS Conformance Document, so you can assume these characters exist on almost all handsets.

You call forth these characters from inside a text file by using the & … ; construct of hypertext markup language (HTML). Thus, for example, £ or £ would bring into being the pound sterling symbol £ on the screen of the handset. The details are covered in RFC 2070, "Internationalization of the Hypertext Markup Language."

	!	"	#	$	%	&	'	(	)	*	+	,	-	.	/
0	1	2	3	4	5	6	7	8	9	:	;	<	=	>	?
@	A	B	C	D	E	F	G	H	I	J	K	L	M	N	O
P	Q	R	S	T	U	V	W	X	Y	Z	[	\	]	^	_
`	a	b	c	d	e	f	g	h	i	j	k	l	m	n	o
p	q	r	s	t	u	v	w	x	y	z	{	|	}	~	□
□	□	‚	ƒ	„	…	†	‡	ˆ	‰	Š	‹	Œ	□	□	□
□	‘	’	“	”	•	–	—	˜	™	š	›	œ	□	□	Ÿ
	¡	¢	£	¤	¥	¦	§	¨	©	ª	«	¬	-	®	¯
°	±	²	³	´	µ	¶	·	¸	¹	º	»	¼	½	¾	¿
À	Á	Â	Ã	Ä	Å	Æ	Ç	È	É	Ê	Ë	Ì	Í	Î	Ï
Ð	Ñ	Ò	Ó	Ô	Õ	Ö	×	Ø	Ù	Ú	Û	Ü	Ý	Þ	ß
à	á	â	ã	ä	å	æ	ç	è	é	ê	ë	ì	í	î	ï
ð	ñ	ò	ó	ô	õ	ö	÷	ø	ù	ú	û	ü	ý	þ	ÿ

Figure 5.1
ISO 8859-1 (Latin 1) character set.

Audio

The thorniest issue to deal with in MMS construction is audio. This may seem strange because an MMS handset is first and foremost a mobile phone.

Most MMS handsets support the AMR format because that is what they use to behave as a mobile phone. AMR is the voice-encoding format used in GSM and 3GPP systems. It is beautiful technology, elegant mathematics, and with all of the pressure on spectrum, highly optimized. If you like standards, read 3GPP TS 26.090, "Mandatory Speech Codec Speech Processing Functions: AMR Speech Codec Transcoding Functions" to learn about AMR. If you are more comfortable reading code than running text, check out 3GPP TS 26.073, "AMR Speech Codec: C-source code."

A number of other 3GPP documents, many referenced by both TS 26.090 and TS 26.073, describe AMR encoding and provide impressive benchmarks of its effectiveness. AMR is the pride of the global system for mobile communication (GSM) system, and justifiably so.

A good introduction to the AMR file format is found in the Internet draft, draft-ietf-avt-rtp-amr-10.txt, "RTP payload format and file storage format for AMR and AMR-WB audio."

Because it is used to digitize human speech, it has been optimized for the human voice and the human vocal tract. If this is what accompanies your MMS messages, you are in good shape. Figure 5.2 shows the frequency response of AMR at various bit rates taken from ETSI TS 101 714 (GSM 06.75), "Performance Characterization of the GSM Adaptive Multi-Rate (AMR) Speech Codec."

Most of the audio formats you are used to dealing with—WAV, AU, MPEG, etc.—have been optimized for the human ear rather than the human mouth. Because the human mouth is different than the human ear (at least in the laboratory humans that were used to do this optimization), you are bound to lose something in the translation.

A number of available tools can get audio from your favorite audio format to AMR. A freeware WAV-to-AMR converter on the Web (usually in a zip file called something like AMRConverter) can get you started. It takes in 16-bit WAV files with 8 kHz sampling, and can turn out AMR files at any of the eight AMR sampling frequencies (see Figure 5.2). The authors have been unable to establish the

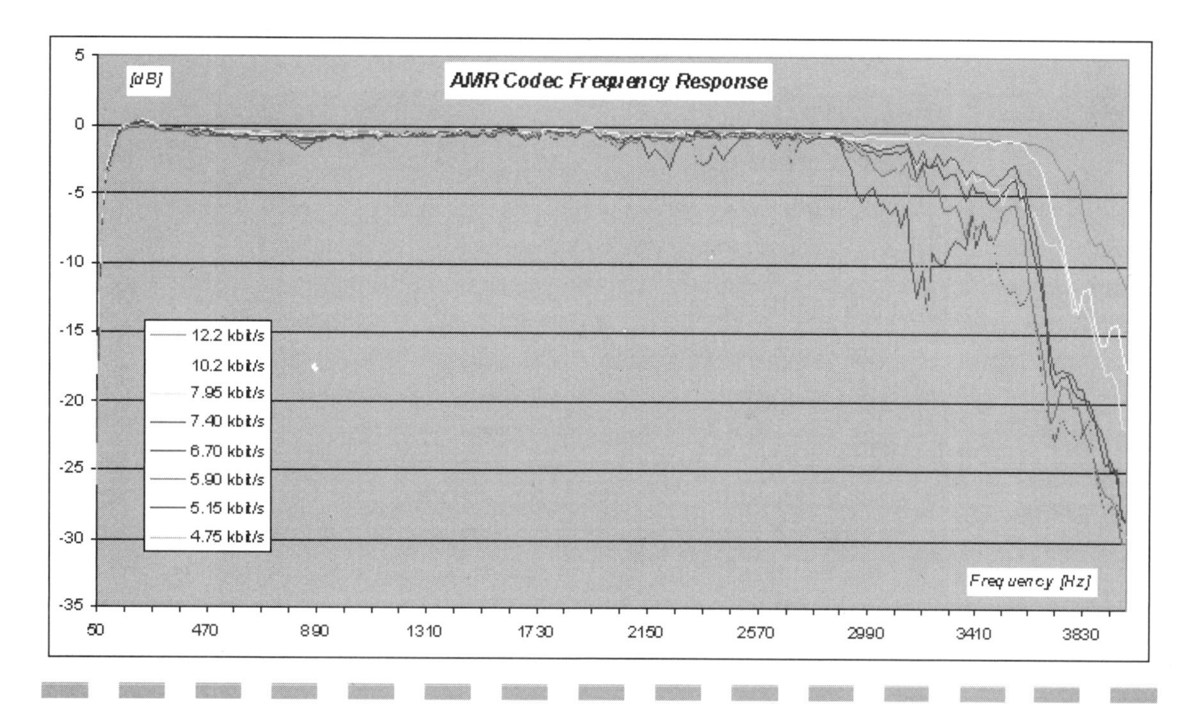

Figure 5.2 Frequency response of AMR encoding.

provenance of the AMRConverter program in order to give its creator(s) due credit. Nokia is also making available an HMR converter called the Nokia Multimedia Converter. You can find it at http://forums.nokia.com.

Table 5.7 is an example of the results of converting a small WAV file to AMR using AMRConverter. You can see that AMR is indeed a very efficient way to represent audio, and perfect for sending audio bits through a congested air space, which is exactly what it was designed to do.

A number of commercial products translate from a number of PC audio formats to and from AMR. The product from VoiceAge called SPOT-XDE Pro sells for around $250 and translates video, too.

VoiceAge has applied for patents on AMR. These are listed in Table 5.8. Since they sell SPOT-XDE, one assumes that one can use the output from the program in commercial endeavors, but if your "legal eagles" are sensitive about such matters, you should probably check.

TABLE 5.7
Example of WAV to AMR Conversion

File	Size in Bytes
Original WAV File	40,814
AMR at 12.2 kbit/s	4102
AMR at 10.2 kbit/s	3462
AMR at 7.95 kbit/s	2694
AMR at 7.40 kbit/s	2566
AMR at 6.70 kbit/s	2310
AMR at 5.90 kbit/s	2054
AMR at 5.15 kbit/s	1798
AMR at 4.75 kbit/s	1670

TABLE 5.8
Some AMR Patents and Patent Applications

Patent or Patent Application Number	Registration Authority	Title
WO02430 53A1 2,327,041	Patent Cooperation Treaty Canada	A Method for Indexing Pulse Positions and Signs in Algebraic Codebooks for Efficient Coding of Wideband Signals
PCT/CA99/01010	Patent Cooperation Treaty	Perceptual Weighting Device and Method for Efficient Coding of Wideband Signals
WO0025298A1 PCT/CA99/01008	WIPO Patent Cooperation Treaty	A Method and Device for Adaptive Bandwidth Pitch Search in Coding Wideband Signals
WO0025303A1 PCT/CA99/01009	WIPO Patent Cooperation Treaty	Periodicity Enhancement in Decoding Wideband Signals
WO0025305A1 PCT/CA99/00990	WIPO Patent Cooperation Treaty	High Frequency Content Recovering Method and Device for Oversampled Synthesized Wideband Signal
WO0137264A1 PCT/CA00/01381	WIPO Patent Cooperation Treaty	Gain-Smoothing Amplifier Device and Method in Codecs for Wideband Speech and Audio Signals

AMR should not be confused with the audio format that is called GSM. GSM is linear predictive code called RPE-LPC (Regular Pulse Excited—Linear Predictive Code). GSM, also called GSM 6.10 after the GSM standard that defines it, compresses 160 13-bit samples into 260 bits (or 33 bytes). At 8 kHz sampling, this yields 13.2 kbit/s, so GSM is not as efficient as AMR. A freeware version of GSM called Toast on the Web, and a commercial product is available from Bullet-Proof Software for around $30.

The default CDMA audio codec, QCELP, seems to be able to carry a much broader range of audio beyond human voice than its GSM counterpart, AMR. Qualcomm makes available a free converter from WAV to QCELP.

To get from other audio formats such as MPEG and AVI to AMR, you may have to go by way of WAV. Table 5.9 compares file sizes for some of the formats you may encounter. It illustrates why WAV will not be used in MMS any time soon.

Synthetic audio. An alternative to trying to map real audio into AMR or QCELP is to use synthetic audio. Initially little more than an upscale doorbell, synthetic audio has come a long way. If it is not a human voice, clear and crisp synthetic audio offers an attractive alternative to clipped and fuzzy real audio.

Besides clarity and volume, synthetic audio is easier to deal with and occupies far, far less space. A MIDI file of an audio track can be between 10 and 50 times smaller than the same audio rendered in AMR. An iMelody file is even smaller.

TABLE 5.9
Comparison of Audio File Size

File	Size in Bytes	Sampling Rate in kHz
WAV	6,723,156	44
MP3 at 320 kbit/s	611,055	48
GSM 6.10 at 13.2 kbit/s	62,094	8
AMR at 12.2 kbit/s	60,966	8
MP3 at 8 kbit/s	38,232	16
AMR at 4.75 kbit/s	24,771	8
MIDI	3191	44

iMelody. iMelody and its predecessors Melody and eMelody have the delightful properties of being pure ASCII, and can drive other output ports on the handset such as the light-emitting diode (LED), the screen background, and the vibrator on the handset.

Both Melody and eMelody are passing out of use because iMelody is so much more agile, while occupying no more space. For the sake of completeness, however, below is a Melody file:

```
MELODY:da+dp+daDpppppppda+dp+dadefgA
```

and the same melody as an eMelody file:

```
BEGIN:EMELODY
VERSION:1.0
MELODY:da+dp+daDpppppppda+dp+dadefgA
END:EMELODY
```

An exercise for the reader is to figure out the melody.

iMelody started out as an advanced ringtone format, but it has grown into an audio format in its own right, with a surprisingly large following. iMelody was invented by the Infrared Data Association and the full specification is available at its Web site, http://www.irda.org. iMelody ringtones can be loaded into handsets using the IrDA link, which is why the format is promoted by this organization.

Below is the classical melody "Bolero" rendered in iMelody:

```
BEGIN:IMELODY
VERSION:1.2
FORMAT:CLASS1.0
MELODY:*4c2c3*3b4*4c4d4c4*3b4a4*4c3c4*3a4*4c2c3*3b4*4c4*3a4g4e4
END:IMELODY
```

A very handy freeware iMelody composer is found at:http://www.christersson.org/software_composer.asp that lets you compose and listen to Melody, eMelody, and iMelody scripts.

No discussion of iMelody would be complete without mentioning the famous walking phone script for the Sony Ericsson T68i handset:

```
BEGIN:IMELODY
VERSION:1.2
FORMAT:CLASS1.0
NAME:Move-A-Phone v1.0
BEAT:200
STYLE:S1
Melody:(vibeonr5vibeoff@100)*0c3*0c3(ledonr3ledoff@10)*0e3*0e3
                        (vibeonr1vibeoff@30)(vibeonr5vibeoff@100)
END:IMELODY
```

Ringtones. Ringtones generate nearly $2 billion in annual revenue for the network operators and music licensing organizations today. Over 150 million ringtones are downloaded every month, and some pundits believe that ringtones could make up 10 percent of the music industry's revenue in a couple of years. There are even companies that sell off-the-shelf, ready-to-go ringtone Web sites.

It is not clear if there is any relationship at all between the market for ringtones (and their graphic tagalongs, phone logos, and screen savers) and the market for MMS. Is a ringtone about listening to music or is it more setting a mood for an incoming call or SMS? Is a ringtone an entertainment or a performance?

There are two major ringtone formats: ringing tones text transfer language (RTTTL) and Nokia Composer ringtone format. RTTTL is sometimes called the Nokring format after the original composer program. Some phones support MIDI for ringtones. Neither RTTTL nor the Nokia Composer ringtone formats are included in the list of MMS supported formats. The reason for this may be simply to keep these two markets apart. Nevertheless, you can send these ringtones via MMS using the application/vnd.nokia.ringing-tone content type.

The new polyphonic ringtones come in two formats, Beatnik RMF file format and SP-MIDI format. SP-MIDI is already supported as an MMS media format in some handsets, and RMF is in the running. These two formats, therefore, represent an overlap between ringtones and MMS. No handset, to the author's knowledge, allows you to associate the playback of an MMS with a handset event such as an incoming call, but it must have occurred to someone.

MIDI and SP-MIDI. A good compromise between full audio such as AMR and MPEG, and ringtone audio such as iMelody, is musical instrument digital interface (MIDI). MMS handsets support a flavor of MIDI called scalable polyphony MIDI (SP-MIDI), which is essentially MIDI with a static sound bank and an added feature where composers can say what notes can be omitted by reduced functionality devices. SP-MIDI files can be a little larger than MIDI files, but not much bigger. A good general overview of MIDI is available at the URL http://www.midi.org/about-midi/gm/gm1sound.htm. A nice little freeware program that converts plain old MIDI files to SP-MIDI files is found at http://www.mmsguru.com.

Graphics

GIF and JPEG. When you think of digital graphics, you think immediately of graphic interchange format (GIF) and Joint Photographic Experts Group (JPEG) files. Both of these graphics formats are supported by most MMS handsets, and GIF, including animated GIF, is in the Conformance Specification.

In general, the file containing a JPEG encoding of a picture is smaller than a file containing a GIF encoding, unless only a very few colors are used in the picture. The screen of existing mobile devices is so small that it is unlikely that any difference in rendering speed between the two formats will be noticeable. On the other hand, animated GIFs do seem to render noticeably slower on handsets compared to what people are used to on their desktops and laptops, so some experimentation is wise if you plan to use animated GIFs in your application.

WBMP and PNG. Wireless bitmap (WBMP) was designed to be an efficient graphics format, but this efficiency is achieved primarily by throwing away color. WBMP is monochromatic. Each pixel in a WBMP graphic is either on or off. On some screens, this will be black and white, and on other screens, this will be green and black. As a one-pixel deep, black-and-white graphics format, WBMP is not particularly efficient or compact. It could have learned a lot from fax machines. WBMP is supported on a number of handsets, however, and therefore is a candidate for encoding where the pictures are freehand black-and-white line drawings. WBMP is described in "Wireless Application Environment Defined Media Type Specification," Version 15-May-2001 (WAP-237-WAEMT-20010515-a).

The color follow-on to WBMP, and some would say to GIF and JPEG too, is the portable network graphics (PNG) format. PNG is described in IETF RFC 2083. PNG offers the quality of JPEG and JPEG2000, along with some of the features of GIF, such as transparency. PNG includes support for gamma and chromacity correction, and contains an RGB-based color management system, which makes it ideal for adapting MMS messages on-the-fly to particular handsets. PNG is not, however, a particularly efficient format for wireless transmission.

Table 5.10 compares the file size of various encodings of a 101 × 59 pixel image that would fill the screen of the Sony Ericsson T68i.

TABLE 5.10 File Size Comparison for Various Raster Graphics Encodings

Encoding	Bits/Pixel	Image Quality	File Size (bytes)	Comment
PNG	24	n/a	6435	Full Color
GIF	8	n/a	5051	Full Color
PNG	8	n/a	4563	Full Color
JPEG	24	80%	1630	Full Color
JPEG	8	80%	1550	Gray Scale
JPEG	24	40%	1153	Full Color
JPEG	8	40%	943	Gray Scale
WBMP	1	n/a	771	Black/White

SVG. Scalable vector graphics (SVG) is to graphics what iMelody is to audio. Like iMelody, SVG is a text-based language for describing a media object, rather than an encoding of a rendering of the object. Unlike iMelody, and indicative of its more recent provenance, SVG is XML-based.

The specification for SVG is maintained by the W3C. Two mobile profiles for SVG have been defined: SVG Tiny and SVG Basic. As with PNG, a key advantage of SVG in the MMS setting is that it can be easily and readily adapted on-the-fly to different handsets. That is the "scalable" part.

A diagonal red line across a 101 × 80 screen (the size of a Sony Ericsson T68i display) in SVG follows:

```
<svg height="80" width="101" >
    <line x1="0" y1="0" x2="101" y2="80" stroke="red"/>
</svg>
```

When this SVG graphic is mapped onto the 176 × 208 screen, horizontal coordinates and distances are multiplied by 176/80, and vertical coordinates and distances are multiplied by 208/101, so the rendered SVG script is:

```
<svg height="176" width="208" >
    <line x1="0" y1="0" x2="208" y2="176" stroke="red"/>
</svg>
```

In other words, it is still a diagonal red line from the upper left hand corner to the lower right hand corner. SVG is very transcoding friendly.

In addition to 2-D Visio-style drawing constructs, SVG supports animation. For example, an animated SVG graphic of a red box drifting from the upper left corner of the display to the lower right corner of the display is shown below:

```
<svg height="80" width="101" >
    <rect width="10" height="10" fill="red">
        <animate attributeName="x" from="0" to="101" dur="10s"
                    repeatCount="indefinite"/>
        <animate attributeName="y" from="0" to="80" dur="10s"
                    repeatCount="indefinite"/>
    </rect>
</svg>
```

Adobe Illustrator supports SVG and distributes browser viewer plug-ins for a number of platforms. There is also a freeware viewer in Java at IBM's Alphaworks. Virtual Mechanics distributes a delightful freeware SVG editor called Web Dwarf, but it does not do animation. Finally, an excellent viewer and SVG library called "Beatnik" is available at xml.apache.org.

As of mid-2003, SVG is not supported in any MMS handsets. A number of companies have come on the scene with handset SVG players, including ZOOMON and BitFlash. MMS handsets and messages harnessing SVG will be eye-popping when and if they arrive, and will provide application developers with a lot of possibilities that are simply not available with GIF and JPEG.

Animation. The only way of realizing animation in an MMS today is with animated GIFs using the GIF89a format. Two other ways are coming along: SVG and synthetic video.

The disadvantage of animated GIFs is that, like a magic lantern or the early cartoons, the entire picture has to be repeated from frame to frame. Furthermore, while in theory GIF89a lets you set the timing in 1/100ths of a second, the number of processor cycles that handsets actually provide to the display of an animated GIF often is not large enough to move from one frame to the next this quickly. Nevertheless, with careful thought about the image itself and image-to-image relationships, surprisingly nice animated MMS messages can be created.

3GPP SMIL does not support mediaRepeat="strip" so you cannot move take the native timing out of the animated GIF and move it into the SMIL script and do the timing there. Synchronizing the animated GIF's timing and the SMIL script's timing is left as an exercise for the programmer.

One of the advantages of SVG animation over GIF89a animation is that you can synchronize these clocks. As shown previously, SVG is an extensible markup language (XML) dialect just like SMIL, so an SVG image inside an SMIL script is just a big tangle of XML. What makes the coordination even more compelling is that SVG animation is very closely modeled after SMIL 2.0 animation. Because both the syntax and semantics of SMIL 2.0 animation and SVG animation are so close, it can be difficult to determine where one stops and the other starts.

While it offers some improvement over the mindless copying all the bits of a picture method of GIF89a, SVG animations can still be large files. Furthermore, SVG provides no 3-D capabilities and thus cannot deliver photorealistic images. For this you have to move to synthetic video.

A number of proposals for incorporating synthetic video into MMS messages have been made to the 3GPP standards groups, including

Flash from MacroMedia and VIM from Vimatix. The VIM format has some very attractive properties, including a very compact representation that is appropriate for low-bandwidth wireless channels, such as those which MMS travels. MMS has to generate a lot of revenue before there will be sufficient pull on the handset manufacturers to include synthetic video in their products.

Miscellaneous

Two miscellaneous media formats presumably could be included currently as one of the many components of an MMS message, but most likely will be the only element in the MMS that carries them: vCard and vCalendar.

A vCard contains information that might be found on a business card and a vCalendar contains information that might be found in an appointment book. The specifications for both of these media formats are maintained by the Internet Mail Consortium (www.imc.org). vCard is also described in IETF RFC 2426 and vCalendar is described in RFC 2445.

Like iMelody and SVG, vCard and vCalendar are text formats. Following is the vCard one of the two authors of RFC 2426 has taken directly from that document:

```
BEGIN:vCard
VERSION:3.0
FN:Frank Dawson
ORG:Lotus Development Corporation
ADR;TYPE=WORK,POSTAL,PARCEL:;;6544 Battleford
Drive;Raleigh;NC;27613-3502;U.S.A.
TEL;TYPE=VOICE,MSG,WORK:+1-919-676-9515
TEL;TYPE=FAX,WORK:+1-919-676-9564
EMAIL;TYPE=INTERNET,PREF:Frank_Dawson@Lotus.com
EMAIL;TYPE=INTERNET:fdawson@earthlink.net
URL:http://home.earthlink.net/~fdawson
END:vCard
```

and below is an example of a vCalendar:

```
BEGIN:VCALENDAR
PRODID:-//ACME/DesktopCalendar//EN
METHOD:REQUEST
VERSION:2.0
BEGIN:VEVENT
ORGANIZER:mailto:sman@netscape.com
ATTENDEE;ROLE=CHAIR;PARTSTAT=ACCEPTED:mailto:sman@netscape.com
ATTENDEE;RSVP=TRUE:mailto:stevesil@microsoft.com
DTSTAMP:19970611T190000Z
DTSTART:19970701T210000Z
DTEND:19970701T230000Z
SUMMARY:Phone Conference
DESCRIPTION:Please review the attached document.
UID:calsvr.example.com-873970198738777
ATTACH:ftp://ftp.bar.com/pub/docs/foo.doc
STATUS:CONFIRMED
END:VEVENT
END:VCALENDAR
```

Handsets are not in total agreement as to what content types to use with vCards and vCalendars. You may have to experiment a little. Some use the text/vCard content type for vCalendar on the grounds that the content of the text file says what it really is. Consult the handset's UAProf if you can.

Summary

It may be said that the success or failure of your MMS application depends in no small part on the media format decisions you make. The recipient of your message will be asking two critical questions:

1. Did I understand or enjoy the message?
2. Was what I understood or enjoyed worth what I paid?

Your choice and organization of the content of the MMS message affects the recipient's answers to these questions of course. But if the graphics come out muddy and the audio is filled with background hiss, it will be difficult to answer the first question positively. On the other hand, a simple message presented with full-color pictures from the Louvre and accompanied by a symphonic audio track will impact the answer to the second question.

The commercial MMS application builder works in engineering as well as design trade-offs. What can be done with the technology is only a starting point. Plan to build many different versions of your MMS messages during development, and carefully measure the properties of each version. How big is it? How does it look on a range of handsets? What is the network operator going to do with it? How do end users react? Only by compiling answers to these questions for your particular application can you figure out what are the best media formats to use in your MMS message.

CHAPTER 6

Testing Your MMS Application

Seeing a multimedia messaging service (MMS) application run on the handset is the ultimate test for any developer. Extensible markup language (XML) editors, synchronized multimedia integration language (SMIL) playback programs, browsers, and handset emulators are fine for roughing out an application, but in the author's experience, these tools do not substitute for seeing your MMS application on a real phone. In fact, these surrogates can lead you astray. For serious MMS application development, plan to test on real handsets.

Does the handset accept your MMS happily and present it accurately? Do the graphics fit on the screen properly? Do the colors come out the way you want them? Is the text legible, in the right place, and in the right font? Are the anchors handy? Do you have the timing right? Do you like how your MMS interacts with the MMS controls on the handset? Do you like the answers to all of these questions on all of the handsets that your customers will be using?

Regardless of what the stack of standards and technical documents next to your personal computer (PC) says about interoperability and compatibility, you should resign yourself to the necessity of sending your MMS to every possible handset to see how it actually behaves "where the rubber meets the road," that is, in your customer's handset. Not only do handsets from different vendors behave differently, but different models from the same vendor behave differently. Surprisingly, even different software releases of the same model from the same vendor can behave differently, and the same software release from the same vendor can behave differently, when bought through different network operators.

You clearly cannot plan, let alone afford, to obtain of every software release of every handset model from every manufacturer. Some handsets you will have in-house, and you may have some friends and friendly customers who own others. Plan to alpha test using the in-house handsets. Beta testing can be done with your friends. Ideally, you will do this frequently during the development of the MMS application. With regular testing, you will not have to undo a month's worth of work because you made some assumptions that turn out to be unwarranted for some of the target handsets.

Getting an MMS onto a handset when you are still in application development mode is the core problem in testing a new MMS application. The following is a list of possible methods you can use to get an MMS onto a handset, in order of decreasing cost and increasing convenience:

1. Upload your MMS to an MMS portal or Web-based MMS composer and send it to yourself over the network.
2. Join an MMS application developer's program and use the program's gateways and service centers to send yourself the MMS.
3. Download the MMS from your own server.
4. Load the MMS through one of the handset's COM ports.

MMS Portals and Web-Based MMS Composers

Some of the online composers are easy to use and support the uploading of your own content. However, there are a number of disadvantages.

First is the turnaround time. Every time you make a change in your application, you have to upload the changed elements and reassemble them using the portal composer, then send the result to yourself.

Next is downloading all the pieces, including the SMIL script, once you have things the way you want them. The business model of most of these portals is to charge for sending an MMS, not to serve as a test platform. Downloading the full code of your completed application back to your development station may prove to be more difficult than uploading and sending the MMS to the handset.

The public context of a portal also presents a potential problem of security and intellectual property protection. You may find yourself competing with the portal, or one of the portal's other customers, for the application you are developing. Uploading your application exposes your ideas and your approach to the application to others before you may want to do so. If you are working with third-party content, the owners of that content may not want it stored on computers over which they have no contractual control.

Last is the issue of cost. Not only will you be expected to cover the full cost of sending each MMS to yourself, the portal will probably add some fees, handling charges, and markups over and above what the network operators are charging.

MMS portals and Web-based composers were not designed to be application development tools, so it is not a surprise that this approach presents a number of problems to the developer. Nevertheless, for certain types of applications and content, it may be worth considering.

MMS Application Developers' Programs

Some network operators and some MMS equipment firms working with network operators offer application developers' programs that include convenient and low-cost access to live network servers that you can use to send MMS messages to yourself.

Some of these programs have the follow-on advantage of making your application's transition from a testing connection to a production connection easier. Furthermore, since the businesses running these programs are trying to stimulate the creation of MMS applications, you may find that you have a built-in sales and distribution channel and a good business partner when you are ready to commercialize your application.

These live developers' program connections are not likely to be free, but they typically cost less than what you would pay to an MMS portal, making the time and effort to find out if there is one that suits your purpose well spent.

Direct Loading of an MMS into the Handset

Some handsets let you directly load an MMS using a local data communication port like the serial connector, the IrDA port, or Bluetooth transmission. A direct connection is the easiest, cheapest, and best way to go. For handsets that support direct loading of an MMS, this is the recommended testing process.

Some handset manufacturers—Sony Ericsson for example—do not support this kind of direct MMS loading on standard versions of their handsets, but they do make special developers' versions of their handsets for this purpose. Obtaining these developers' specials in a timely fashion can sometimes be a challenge.

Since using a handset to send an MMS is what the MM1 interface is all about, it is a puzzle that handset manufacturers and network operators don't make it as easy as possible to move an MMS from a PC onto a phone, and then out onto the network. Maybe this is in the works, but for right now, this way of testing your MMS application is hard to implement.

Sending an MMS from Your Own Server

While it is neither as convenient nor as cheap as directly loading the MMS into a handset using a handset communications port, sending an MMS to the handset from your own server is still cheaper, more convenient, and more secure than using an MMS portal or a Web-based composer to send yourself each test MMS.

You still have to pay the freight for how you access your server, whether it is a circuit-switched dial-up or HSCSD connection, or a packet-switched GPRS or EDGE connection. However, on a per-byte basis, these connections are typically cheaper than what it costs to send an MMS. And, they have the additional advantage that the operator does not perform some unknown and uncontrollable transformations on your MMS. Bypassing the operator allows you to see the MMS the way you are building it, rather than the way the operator thinks it ought to look.

This approach has the additional advantage that you learn a lot about the bits and bytes of MMS messages, and as a result, you can grab hold of them and make them do what you want if that becomes necessary.

ASCII versus Binary Representation of MMS Messages

Almost all the technical documentation about MMS, including all the MMS standards, describes MMS messages in their readable ASCII representation. Thus, for example, an MMS message arriving at a handset is presented for the purposes of reading and understanding as:

```
X-Mms-Message-Type: m-retrieve-conf
X-Mms-Transaction-ID: 777
X-Mms-MMS-Version: 1.0
To: +16177813540009/TYPE=PLMN
From: sguthery@mobile-mind.com
Date: 27 November 2002
Content-Type: application/vnd.wap.multipart.related;
start=<A0>;
boundary="boundary123456789";
type=application/smil
--boundaryq123456789
Content-ID: <A0>
Content-Type: application/smil; charset="US-ASCII"
<smil>
[. . .]
</smil>
--boundary123456789
Content-Location: q/Text.txt
Content-Type: text/plain
This is the text of the second slide.
--boundary123456789
Content-Location: q/Sound.amr
Content-Type: audio/AMR
[. . .]
--boundary123456789
Content-Location: q/Image.jpg
Content-Type: image/jpeg
[. . .]
--boundary123456789-
```

This ASCII representation of the MMS is not sent over the air from the multimedia messaging service center (MMSC) to the handset, and it certainly is not what a handset is programmed to digest. Instead, the handset wants a compact binary encoding of the message. The binary encoding, not the above ASCII description, is sent over the air. In loading the handset—either directly using a COM port or from your own server—give the handset what it wants, not what we use to write and talk about MMS messages.

The two most important envelopes for loading an MMS onto a handset are m-notification-ind and m-retrieve-conf. The first tells the handset that an MMS message is available for retrieval, and the second is the response to a retrieval request; that is, the MMS itself. In normal MMS network operation, the MMSC pushes the notification to the mobile using a short messaging service (SMS) or a wireless application protocol (WAP) push, and then the mobile sends a retrieval request back to the MMSC to get the MMS message.

The Binary Representation of the "Hello, World" MMS Notification

The following is what the m-notification-ind message for our "Hello, world" MMS looks like in human readable form:

```
X-Mms-Message-Type: m-notification-ind
X-Mms-Transaction-ID: 0000000001
X-Mms-MMS-Version: 1.0
X-Mms-Sender-Address: sguthery@mobile-mind.com
X-Mms-Subject: Hello
X-Mms-Message-Class: Informational
X-Mms-Message-Size: 396
X-Mms-Expiry: 12:00:00
X-Mms-Content-Location: http://mms.mobile-mind.com/q/hello.mms
```

To get an MMS handset to swallow this notification message, we must encode it according to the WAP MMS encapsulation protocol. Roughly, each header field name and each fixed-choice header field value turns into a single byte, and each text string turns into … well, a text string.

The complete binary m-notification-ind envelope, just as it is pushed from the MMSC to your mobile phone follows:

```
8C8298303030303030303031008D90891A80736775746865727940 6D6F62696
C652D6D696E642E636F6D009648656C6C6F008A828E018C88048102A8C0836874
74703A2F2F6D6D732E6D6F62696C652D6D696E642E636F6D2F672F68656C6C6F2
E6D6D7300
```

Let's decode this MMS encapsulation protocol encoded notification envelope:

```
8C = Message-Type
82 = m-notification-ind

98 = Transaction ID
30303030303030303100 = null-terminated ASCII string
0000000001

8D = MMS Version
90 = 1 001 0000 = 1.0 (special encoding)

89 = From
1A = Length of following data
80 = Address present token
736775746865727940 6D6F62696C652D6D696E642E636F6D00 = null-
  terminated ASCII string
sguthery@mobile-mind.com

96 = Subject
48656C6C6F00 = null-terminated ASCII string
Hello

8A = Message-Class
82 = Informational

8E = Message Size
02 = Length of following integer in bytes
```

```
018C = Number of bytes in the MMS message (396)

88 = Expiry
04 = Length of following data
81 = Relative time in seconds
02 = Length of integer seconds value in bytes
A8C0 = Hold message for 43200 seconds (12 hours) on the MMSC

83 = URL of Content Location
687474703A2F2F6D6D732E6D6F62696C652D6D696E642E636F6D2F672F68656C6
C6F2E6D6D7300 = null-terminated ASCII string
http://mms.mobile-mind.com/q/hello.mms
```

All of these encodings can be found in two Open Mobile Alliance documents. WAP-209-MMS Encapsulation-20020105-a gives the MMS-specific details of this encoding, such as the fact that the Message Class header field name is encoded as the single byte 0x8A. WAP-230-WSP, the specification for the WAP wireless session protocol (WSP), gives the details of the generic elements, such as how various integer and string values are represented.

You may also need to visit the Wireless Interim Naming Authority (WINA) at http://www.wapforum.org/wina/ for late breaking header and content type encodings. WINA is the Open Mobile version of the Internet Association Naming Authority (IANA); that is, the keeper of the assigned numbers.

We have now reduced the problem of getting an MMS on a handset to getting an MMS notification message to the handset, or more specifically, the MMS application or subsystem of the handset. If we can get the notification message to the handset's MMS application, it will use the URL of the content location in the notification message to pull down the MMS itself.

We will describe the following three ways of getting an MMS notification message to a handset:

1. As a wireless markup language (WML) page
2. As an SMS message.
3. As a WAP push.

An MMS Notification as a WML Page

Getting the notification message to the handset as a WML page is the easiest and fastest way to go. It is also the cheapest—a good thing from your point of view, but a bad thing from the network operator's point of view. The result is that starting at the end of 2002, many MMS handsets have been programmed to not allow this. Since a lot of early handsets do, we will cover this approach anyway. The unintended consequence of blocking this way of getting an MMS notification may be to increase the value of these old handsets.

To get a notification message onto a handset as a WML page, create another WML page that has an anchor pointing to a binary file containing the above WSP-encoded notification message. To get the binary data in the file passed to the MMS subsystem of your handset, make sure the WAP server serves this file with the multipurpose Internet mail extension (MIME) type of application/vnd.wap.mms-message.

To accomplish this, configure the WAP server to associate all files ending in the file extension ".mms" with this MIME type.

The WML page to get this file to your handset might look something like the following:

```
<?xml version="1.0"?>
<!DOCTYPE wml PUBLIC "-//WAPFORUM//DTD WML 1.1//EN"
    "http://www.wapforum.org/DTD/wml_1.1.xml">
<wml>
  <head>
  <meta http-equiv="Cache-Control" content="max-age=30"
   forua="true" />
  </head>
  <card id="card0" title="Mobile-Mind">
    <do type="prev" label="Back"><prev/></do>
    <a href="http://mms.mobile-mind.com/notification.mms">Hello
     MMS</a>
  </card>
</wml>
```

This file might be named something like mmsnotification.wml. It is a regular .wml page, not a binary .mms page.

When you start the WAP browser on your handset, point at this WML page, and click on "Hello MMS," the binary file notification.mms is sent to your handset via the WAP gateway. Because this incoming message has MIME type application/vnd.wap.mms-message, the WAP browser on your handset knows to pass it to the MMS application. The MMS application code looks at the first two bytes of the message, sees that it is an MMS notification message, and puts it into the MMS in-box on your handset. Bingo!

Of course, if we were in production mode rather than development mode, the notification.mms bytes would have been WAP-pushed to your handset by the MMSC rather than your having to go to the MMSC to see if there were any incoming MMS messages waiting for you. Regardless of whether they are pushed or pulled, the bytes of the notification message handed to the MMS application of your handset would be exactly those described previously.

From this point forward, our development process is similar to what happens with real MMS messages. Clicking on a notification message in the MMS in-box causes the Content-Location URL in the notification message to be sent back to the WAP gateway as an m-retrieve-req envelope. The WAP gateway fetches the MMS message from the specified location and sends it back to the handset as the response to the handset's m-retrieve-req envelope; that is, as an m-retrieve-conf envelope.

Similar to the notification message, the m-retrieve-req envelope will have MIME type application/vnd.wap.mms-message, so it will be handed over to the MMS application on the handset. The only difference is that this time, it is a real MMS message, not just a "You've got an MMS" notification message.

The Binary Representation of the "Hello, World" MMS Message

Before looking at the other ways of getting a notification message to the handset, let's take a little detour to look at the MMS encapsulation protocol binary encoding of the MMS message itself. The day may come when you have to pick through the bits and bytes to figure out what is happening.

The following is how the m-retrieve-conf envelope containing our example "Hello, world" MMS message looks in the human readable form that you will see in the MMS specifications and literature:

```
X-Mms-Message-Type: m-retrieve-conf
X-Mms-Transaction-ID: 0000000001
X-Mms-MMS-Version: 1.0
X-Mms-Date: Saturday, 23 Nov 2002 16:23:32 GMT
X-Mms-Sender-Address: sguthery@mobile-mind.com
X-Mms-Subject: Hello, world
X-Mms-Content-Type: application/vnd.wap.multipart.related;
  start=<A0>;
                    boundary="boundary123456789";
type=application/smil

--boundary123456789
Content-Location: hello.txt
Content-Type: text/plain; charset="US-ASCII"
Hello, world

--boundary123456789
Content-ID: <A0>
Content-Type: application/smil; charset="US-ASCII"

<smil>
  <head>
    <layout>
      <region id="text" top="0" left="0" height="100%"
       width="100%"/>
  </layout>
  </head>
  <body>
    <par dur="5000ms">
      <text src="hello.txt" region="text" />
    </par>
  </body>
</smil>
```

Below is the MMS encapsulation protocol-encoded envelope that MMSC actually sends to the handset:

```
8C849830313233343536373839008D9085043DDFA8B4891A80736775746865727
9406D6F62696C652D6D696E642E636F6D009648656C6C6F2C20576F726C640084
19B3896170706C69636174696F6E2F736D696C008A3C41303E00020F0E0383818
38E68656C6C6F2E7478740048656C6C6F2C20776F726C640D0A1B816D13617070
6C69636174696F6E2F736D696C008183C0223C41303E003C736D696C3E0D0A202
03C686561643E0D0A202020203C6C61796F75743E0D0A20202020202020203C726567
696F6E2069643D2274657874222220746F703D223022206C6566743D2230222220686
5696768743D223130302522207769647468 3D223130302522F3E0D0A20203C2F
6C61796F75743E0D0A20203C2F686561643E0D0A20203C626F64793E0D0A20202
0203C706172206475723D22353030306D73223E0D0A20202020202020203C74657874
207372633D2268656C6C6F2E7478742220726567696F6E3D227465787422202F3
E0D0A202020203C2F7061723E0D0A20203C2F626F64793E0D0A3C2F736D696C3E
```

Let's dissect this MMS encapsulation protocol envelope as we did with the m-notification-req envelope and make sure we get back to the human readable form. Keep in mind that all the single-byte values are encoded as WSP short integers, which means that the most significant bit is set. For example, 0x0C is encoded as 0x8C. To get the value that is found in the MMS and WSP documents, mask off the top bit:

```
8C = Message-Type
84 = m-retrieve-conf

98 = Transaction ID
30303030303030303031 00 = null-terminated ASCII string
0000000001

8D = MMS Version
90 = 1 001 0000 = 1.0 (special encoding)

85 = Date
04 = Number of bytes in following integer
3DDFA8B4 = Number of seconds from 1970-01-01, 00:00:00 GMT
          (1,038,067,892)

89 = From
1A = Length of following data
80 = address present token
7367757468657279406D6F62696C652D6D696E642E636F6D00 = null-
      terminated ASCII string
```

```
sguthery@mobile-mind.com

96 = Subject
48656C6C6F2C20576F726C6400
Hello, World

84 = Content-Type
1B = Content-Type Length
B3 = application/vnd.wap.multipart.related

89 = Type Parameter of Content-Type
6170706C69636174696F6E2F736D696C00
application/smil

8A = Start Parameter of Content-Type
3C41303E00
<A0>

02 = Number of Parts in the MIME Body

Start of the First Part
======================
0F = Length of Header of First Part
0E = Length of Data of First Part

Headers of the First Part
-------------------------
03 = Content-Type Length
83 = text/plain
8183 = charset/usascii

8E = Content Location
68656C6C6F2E74787400 = null terminated ASCII string
hello.txt

Data of the First Part
----------------------
48656C6C6F2C20776F726C640D0A

Hello, world\r\n
```

```
Start of the Second Part
========================
1B = Length of Header of Second Part
81 6D = Length of Data in Second Part (237)

Headers of the Second Part
--------------------------
13 = Length of Content-Type
6170706C69636174696F6E2F736D696C00
application/smil

8183 = charset/usascii

C0 = Content-ID
22 = " (quote)
3C41303E00
<A0>

Data of the Second Part
-----------------------
3C736D696C3E0D0A20203C686561643E0D0A202020203C6C61796F75743E0D0A2
020202020203C726567696F6E2069643D22746578742220746F703D223022206C
6566743D223022206865696768743D22313030252220776964746683D22313030
5222F3E0D0A20203C2F6C61796F75743E0D0A20203C2F686561643E0D0A20203C
626F64793E0D0A202020203C706172206475723D22353030306D73223E0D0A202
0202020203C7465787420737263 3D2268656C6C6F2E747874222072656769 6F6E
3D227465787422202F3E0D0A202020203C2F7061723E0D0A20203C2F626F64793
E0D0A3C2F736D696C3E

<smil>
  <head>
    <layout>
      <region id="text" top="0" left="0" height="100%"
width="100%"/>
  </layout>
  </head>
  <body>
    <par dur="5000ms">
      <text src="hello.txt" region="text" />
    </par>
  </body>
</smil>
```

Note that the boundary construct of ASCII MIME encoding has been replaced by byte counts in the WSP multipart data encoding.

An MMS Notification as an SMS Message

If you cannot pull the MMS notification message to your handset, the next best thing is to push it there. You can do this using an SMS message or using a WAP push message. Because SMS has been around longer than WAP, and because there are plenty of SMS software and SMS delivery alternatives, we will look at an SMS push of an MMS notification message first.

The payload of the SMS message will be the binary MMS notification message we tore apart previously, so there is no need to review it. The trick is to get the SMS subsystem on the handset to pass the notification message to the MMS application the same way we got the browser to pass the notification message to the MMS application. In other words, what is SMS talk for a content type of application/vnd.wap.mms-message?

An information element in the User Data Header section of the SMS message is used to achieve this. The particular element used is "Application port addressing scheme, 16 bit address."

The following is the front end of the SMS-SUBMIT transport protocol data unit (TPDU), up to the binary MMS notification message itself that we have to send to our local short messaging service center (SMSC) in order to push the notification message to the MMS application on the handset:

```
41000B916171979291F400046c0605040B8423F000060403BE8183
```

Let's tear this TPDU apart according to 3GPP TS 23.040, "Technical Realization of the Short Message Service" (Release 6).

```
41 = 0100 0001--SMS Flags: User Data Present (0100)
                           and a SMS-SUBMIT message (0001)

00 = Message Reference

0B916171979291F4 = Destination Address--the handset onto which
```

```
                                    we want to get the MMS

00 = Protocol Identifier - implicit-device type is specific to
                           this Service Center, or can be
                           concluded on the basis of the address

F5 = Data Coding Scheme--binary encoding, mobile equipment
                        specific

6c = User Data Length - the length of all the following data;
                        i.e. User Data Headers and User Data
--------------
Start of User Data Headers

06 = User Data Header Length--the total length of the
                               User Data Headers

05 = Information Element Identifier - Application port
                               addressing scheme, 16 bit address

04 = Length of this Information Element

0B84 = Destination Port--2948 = WAP Push
                           (port number from http://www.IANA.com)

23F0 = Origination Port - 9200 = WAP Connectionless Session
                            Service
                           (port number from http://www.IANA.com)

End of User Data Headers
--------------
Start of User Data

00 = Transaction ID

06 = PDU Type (Push)

03 = Length of Content Type and Headers

BE = Content-Type (0x80 + 0x3E) - application/vnd.wap.mms-message
```

```
8183 = Character Set--US ASCII

End of PDU Headers
--------------
Start of MMS Notification Message

8C = Message-Type
82 = m-notification-ind

98 = Transaction ID
3030303030303030303100 = null-terminated ASCII string
0000000001

etc.
```

Note that the content type of the WAP push tucked inside the End of User Data Headers section of the SMS message is application/vnd.wap.mms-message.

This is by no means the only way to encode the SMS push of an MMS notification message. You may have to experiment a little to get it working on your network. The following script is a small C program that you can hitch up to the data cable on your mobile to send out an MMS notification using SMS.

```
#include <windows.h>
#include <stdio.h>
#include <time.h>

typedef unsigned char BYTE;

void sendMMSNotification(char *smsc, char *to, char *from, char
*subject, char *URL);
void hexString(char *h, char *t, char *s);
void hexString80(char *h, char *t, char *s);

#define PRINTERROR FormatMessage(FORMAT_MESSAGE_ALLOCATE_BUFFER
|\
                      FORMAT_MESSAGE_FROM_SYSTEM |\
                                 FORMAT_MESSAGE_IGNORE_INSERTS,\
                                 NULL, GetLastError(),\
                                 MAKELANGID(LANG_NEUTRAL,
```

```
                                        SUBLANG_DEFAULT),\
                                        (LPTSTR)&lpMsgBuf, 0, NULL)

#define sendc(command) \
     bytes = 0; \
     if(WriteFile(handle, command, strlen(command), &bytes, NULL)
             == 0) PRINTERROR;\
     printf("To handset: %s\n", command);\
     Sleep(2000); bytes=0;\
     if(ReadFile(handle, buffer, sizeof(buffer), &bytes, NULL)
              == 0) PRINTERROR;\
     buffer[bytes] = '\0'; printf("From handset[%d]:%s\n", bytes,
              buffer);

HANDLE handle;

int main()
{
    handle = CreateFile("COM1",
                  GENERIC_READ | GENERIC_WRITE,
                  0,      // exclusive access
                  NULL,   // no security
                  OPEN_EXISTING,
                  0,      // no overlapped I/O
                  NULL);  // null template

    sendMMSNotification("12063130004", "16177929194",
          "sguthery@mobile-mind.com", "Hello",
          "http://mms.mobile-mind.com/q/hello.mms");
}

char *CMGSh = "07919171095710F0";     // SMSC Number

char *AT    = "AT\r";
char *ATZ   = "ATZ\r";
char *ATE0  = "ATE0\r";
char *ATE1  = "ATE1\r";
char *ATI0  = "ATI0\r";

char *CMGSo = "4100";    // SMS Flags (UDH Indicator and
      SMS-SUBMIT) and Message Reference
```

```
char *CMGSm = "0004";    // Protocol Identifier and Data Coding
      Scheme
char *CMSGt = "\032\r";

char *UserDataHeader = "0605040B8423F0";
char *WAPPush = "00060403BE8183";
char *MMS1 = "8C82983031008D90";
char *MMS2 = "8A828E0178880680047AE9924A";

void sendMMSNotification(char *smsc, char *to, char *from, char
      *subject, char *URL)
{
    BYTE CMGSc[20], buffer[512], TPDU[512], UDL[4];
    BYTE hSMSC[64], hTo[64], hFrom[64], hSubject[64], hURL[256];
    int i, bytes, n;
    LPTSTR lpMsgBuf;

    strcpy(hSMSC, "07910000000000F0");
    for(i = 0; i < 11; i++)
        hSMSC[5+i-2*(i%2)] = smsc[i];

    strcpy(hTo, "0B910000000000F0");
    for(i = 0; i < 11; i++)
        hTo[5+i-2*(i%2)] = to[i];

    hexString80(hFrom, "89", from);
    hexString(hSubject, "96", subject);
    hexString(hURL, "83", URL);

    n = (strlen(UserDataHeader) + strlen(WAPPush) +
        strlen(MMS1) + strlen(hFrom) + strlen(hSubject) +
        strlen(MMS2) + strlen(hURL))/2;

    /* Initiate the SMS-SUBMIT command */
    sprintf(CMGSc, "AT+CMGS=%02d\r", 13+n);
    sprintf(UDL, "%02x", n);

    /* Build the TPDU */
    strcpy(TPDU, hSMSC);                /* SMSC Number */
    strcat(TPDU, CMGSo);                /* SMS Flags and Message
                                           Reference */
```

```
    strcat(TPDU, hTo);                /* Address of Recipient */
    strcat(TPDU, CMGSm);              /* Protocol Identifier and
                                         Data Coding Scheme */
    strcat(TPDU, UDL);                /* Length of User Data
                                         Headers and User Data */
    strcat(TPDU, UserDataHeader);     /* Destination Port of the
                                         SMS: WAP Push */
    strcat(TPDU, WAPPush);            /* WAP Push PDU */

    /* Add the MMS Notification Message */
    strcat(TPDU, MMS1);               /* Message Type, Transaction
                                         ID, Version */
    strcat(TPDU, hFrom);              /* From */
    strcat(TPDU, hSubject);           /* Subject */
    strcat(TPDU, MMS2);               /* Message Class, Message
                                         Size, Expiry */
    strcat(TPDU, hURL);               /* Location */

    /* Tack on the trailer of the PDU */
    strcat(TPDU, CMSCt);

    sendc(ATE1);
    sendc(AT);
    sendc(ATZ);
    sendc(AT);

    sendc(CMGSc);
    sendc(TPDU);

    sendc(ATE0);
    sendc(AT);
    sendc(ATI0);
}

static BYTE
hexChar (BYTE b)
{
  return (b < 10) ? ((BYTE)'0' + b) : ((BYTE)'A' - 10 + b);
}

void
```

```
hexString(char *h, char *t, char *s)
{
    BYTE l = strlen(s)+1;

    strcpy(h, t);
    h += 2;

    while(*s) {
        *h++ = hexChar((BYTE)((*s)>>4));
        *h++ = hexChar((BYTE)((*s)&0x0F));
        s++;
    }
    *h++ = '0';
    *h++ = '0';

    *h = '\0';
}

void
hexString80(char *h, char *t, char *s)
{
    BYTE l = strlen(s)+2;

    strcpy(h, t);
    h += 2;

    *h++ = hexChar((BYTE)(l>>4));
    *h++ = hexChar((BYTE)(l&0x0F));
    *h++ = '8';
    *h++ = '0';
    while(*s) {
        *h++ = hexChar((BYTE)((*s)>>4));
        *h++ = hexChar((BYTE)((*s)&0x0F));
        s++;
    }
    *h++ = '0';
    *h++ = '0';

    *h = '\0';
}
```

There are also commercial solutions for sending MMS notification messages, such as the Now SMS/MMS Gateway, of which a flexible evaluation version can be downloaded at http://www.nowsms.com.

An MMS Notification as a WAP Push

One form of WAP push is a combination of the notification message download and SMS techniques described previously. You send the phone a WAP push service indication (SI) that contains the URL of the MMS notification file as the indication. If the WAP push bearer you use is SMS, this service indication is sent to your phone as an SMS message. An example of a WAP push SI message to send an MMS notification message to the phone is given in the following code.

When you say "Yes" to the load request of the SI message, the URL of the MMS notification message is sent back to the WAP gateway, which goes to your server, gets the MMS notification file, and sends it to the phone as another SMS message; effectively, the SMS message described previously. Now say "Yes" to the MMS notification message, and the MMS application sends the URL of the MMS message in the MMS notification message back to the WAP gateway, which goes back to your server to get the MMS, and ships it back to the MMS application on the phone. (And the music comes out over here!)

```
Content-Type: multipart/related; boundary=asdlfkjiurwghasf;
      type="application/xml"
--asdlfkjiurwghasf
Content-Type: application/xml

<?xml version="1.0"?>
<!DOCTYPE pap PUBLIC "-//WAPFORUM//DTD PAP 1.0//EN"

http://www.wapforum.org/DTD/pap_1.0.dtd">
<pap product-name="Push Initiator">
<push-message push-id="11" progress-notes-requested="false">
           <address address-value=WAPPUSH=+16177929194/
            TYPE=PLMN@ppg.carrier.com/>
         <quality-of-service delivery-method="unconfirmed"
                    bearer="SMS" bearer-required="true"/>
```

```
</push-message>
</pap>

--asdlfkjiurwghasf
Content-type: text/vnd.wap.si

<?xml version="1.0"?>
<!DOCTYPE si PUBLIC "-//WAPFORUM//DTD SI 1.0//EN"
"http://www.wapforum.org/DTD/si.dtd">
<si>
<indication si-id="1" href="http://mms.mobile-mind.com/hello.mms"
action="signal-high">MMS Notification</indication>
</si>

--asdlfkjiurwghasf--
```

Essentially, we are using the WAP push SI as a substitute for clicking on the MMS notification message anchor on a WML page. A more direct route, of course, is to let the WAP push be the MMS notification message. An example of a direct WAP push of an MMS notification message is given in the following code:

```
Content-Type: multipart/related;
boundary=asdlfkjiurwghasf; type="application/xml"

--asdlfkjiurwghasf
Content-Type: application/xml

<?xml version="1.0"?>
<!DOCTYPE pap PUBLIC "-//WAPFORUM//DTD PAP 1.0//EN"
http://www.wapforum.org/DTD/pap_1.0.dtd">
<pap product-name="Push Initiator">
    <push-message push-id="11" progress-notes-requested="false">
    <address address-value=WAPPUSH=+16177929194/
     TYPE=PLMN@ppg.carrier.com/>
    <quality-of-service delivery-method="unconfirmed"
         bearer="SMS"
         bearer-required="true"/>
    </push-message>
</pap>
```

```
--asdlfkjiurwghasf
Content-type: application/vnd.wap.mms-message
--asdlfkjiurwghasf--

(Note: Here you would insert the binary MMS Encapsulation
Protocol encoding of the text MMS headers displayed below.)

X-WAP-Application-Id: mms.ua
X-Mms-Message-Type: m-notification-ind
X-Mms-Transaction-Id:0
X-Mms-Version: 1.0
X-Mms-Message-Class: Personal
X-Mms-Message-Size: 376
X-Mms-Expiry: 256;type=relative
X-Mms-Content-Location:http://mms.mobile-mind.com/q/hello.mms

--asdlfkjiurwghasf--
```

In actual fact, the WAP push gateway would be given the binary encoding of thc X-Mms headers.

Summary

Testing your MMS application can be accomplished only by getting your MMS messages on real handsets and seeing how they will look and feel to your users. By and large, SMIL players and handset simulators and emulators are not adequate testing devices because they do not convey how your application really appears to the customer.

The biggest challenge in developing your MMS application is tightly integrating handset testing into your application development cycle without blowing your development budget on network charges. We described three ways of doing this: developers' programs, direct loading, and use of your own WAP server.

CHAPTER 7

Using MMS for Mobile Delivery of Content—The MM3 Interface

If multimedia messaging service (MMS) traffic is to even approximate the magnitude of the network operator's dreams, then it is unlikely that most MMS messages will be handcrafted. Sending a picture taken with a phonecam as MMS may be fun on occasion, but sending thousands or millions of pictures is not the work of a casual user. Building an MMS and sending it off from a Web site, or using the MMS construction tools described in Chapter 4, will not scale to mass market levels. These are all labor-intensive activities that are not what most people would do for the joy of it.

If MMS takes off, we can expect that a significant percentage of the MMS messages flowing in the network will be generated automatically from existing content sources. Unlike short messaging service (SMS), where one can generate a message quickly using a mobile handset and a short text message, multipart MMS messages (not just sending a snapshot) take much more time, effort, and careful thought.

Both the MM3 and MM7 interfaces in the MMS architecture are for machine-generated MMSs. MM3 was used by what the 3GPP standards call legacy systems, by which they evidentially mean non-mobile messaging and communication systems, such as e-mail and fax, that were around before MMS existed. MM7 is for new systems that presumably are designed specifically for MMS and mobile applications. In fact, there is little technical distinction between these two interfaces. The difference relates more to the business model that informs them. From a technical point of view, both interfaces are run by volume MMS providers who are information technology (IT) professionals and are probably running the interface as a business endeavor.

This chapter focuses on the MM3 interface. This is where we connect existing protocols to the MMS architecture. It is also the interface on which transcoding, or what is called content adaptation, happens. (There is a move afoot to move this process to a new MM9 interface, but we will not discuss that eventuality here.) Figure 7.1 is a diagram of the expanded view of MM3.

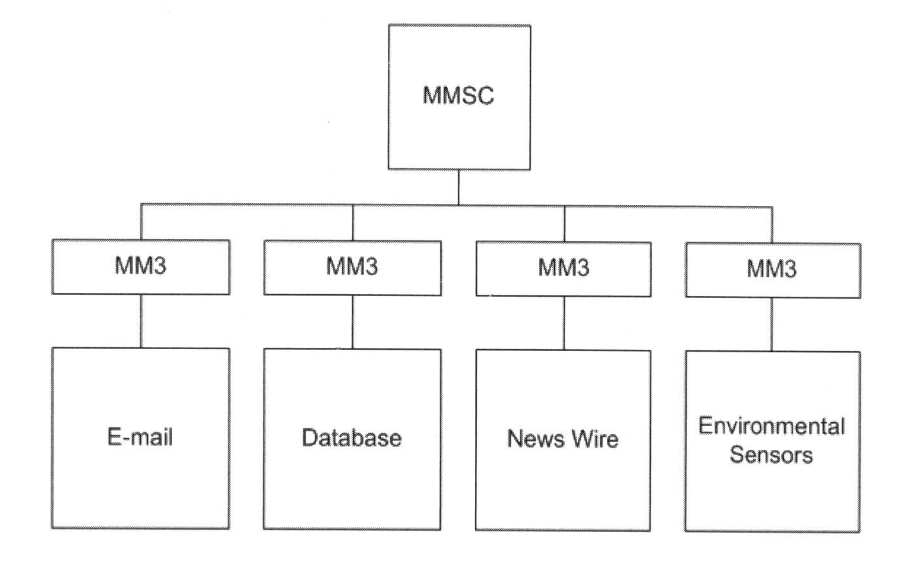

Figure 7.1
Expanded MM3.

E-Mail and MMS

The 900-pound legacy system sitting in the middle of the room is e-mail. Both e-mail and MMS are multipurpose Internet mail extension (MIME)-encoded. A no-brainer, right? Just let it flow. Well, not quite. The formats are essentially the same, but the operational semantics are different. As a result, the design of MMS systems is different from the process of sending and receiving e-mail. Table 7.1 lists some of the differences.

TABLE 7.1
E-Mail versus MMS

	E-Mail	MMS
Basic Operating Paradigm	Supply push	Demand pull
Prior Notification Before Delivery	No	Yes
Cost of Delivery Based on Message Size	No	Yes
Delivery Time	Virtually immediate	Arbitrary delay
Audio Content	Rare	Customary
Delivery Reports	Rare	Customary
Filtering by the Carrier	No	Yes

In a sense, MMS is more polite than e-mail. It tells you that you have a message before laying it on you. Furthermore, it tells you some interesting things about the message, such as who it is from, how big it is, how much it will cost to receive it, and the subject of the message. For many messages we receive, this information is sufficient. Besides taking delivery of the message, we can delete the message, forward the message, or simply note that it has arrived and go pick it up later.

The universal e-mail delivery protocol (as opposed to e-mail encoding protocol) is the simple mail transfer protocol (SMTP; RFC 821). SMTP likes to move things along, so we have to add a notification phase in between SMTP and MMS. Fortunately, the Internet Engineering Task Force (IETF) has defined a couple of protocols that do this. One is called simple notification and alarm protocol (SNAP; Internet Draft draft-shapira-snap-05). Another is called specific event notification (RFC 3265). The latter approach is part of the session initiation protocol (SIP) that is also finding its way into the 3G architecture. Because it is somewhat simpler and more of an independent entity, we will use SNAP in our example, but both perform the same basic function, and either could end up being in the MMS specifications.

There are a number of ways to wire SMTP and SNAP together to make an MM3 interface to an MMSC, and all these alternatives are currently being debated in the 3GPP MMS working groups.

A rough sketch of the incoming e-mail drill will probably look something like this:

1. E-mail arrives at your e-mail account.
2. Your e-mail server sends a SNAP notification to your Multimedia Messaging Service Center (MMSC).
3. The MMSC, after contemplating the SNAP notification, translates it into an MMS m-notifiy-req and sends it to your handset.
4. You decide what you want to do with the e-mail.
   a. Delete it—MMSC sends an SMTP delete command back to your e-mail server.
   b. Forward it—MMSC sends an SMTP forward command back to your e-mail server.
   c. Read it—MMSC retrieves the e-mail using SMTP, translates it into an MMS, and sends it to your handset.

Using the SNAP protocol will likely result in something being put in the m-notifiy.req—a new message class called e-mail, for example, that lets you know this is an MMS-ified e-mail, rather than a real MMS. It also lets you know that the message is not really sitting on the MMSC that sent you the notification, but is sitting on your e-mail server.

The outgoing drill (from the point of view of the MMS system) is even simpler and is already implemented in most MMSCs. The MMSC repackages the outgoing MMS as an e-mail and pushes it to a local e-mail server that is then responsible for getting it to the ultimate e-mail address.

There are many technical details to work out, such as standardizing the translation between MMS and e-mail formats, and what to do with media formats like adaptive multirate (AMR), but they are all technical details. There are also other ways of implementing the notification protocol between the MMSC and the e-mail server, such as the event notification facility in the session initiation protocol (SIP; RFC 3265).

The business details are, and will be, more problematic. For instance, how much does it cost to get an e-mail via MMS and what filtering does the network operator decide to perform?

Once we have an MM3 gateway between e-mail and MMS, other media formats that have been represented in e-mail can ride along for free. For example, we can map fax to and from MMS via our e-mail gateway by using RFC 2305, "A Simple Mode of Facsimile Using Internet Mail" and RFC 2532, "Extended Facsimile Using Internet Mail."

In fact, the technique we use for SMTP can work with almost any messaging protocol including, for example, voice mail. Add SNAP to impedance match between the push semantics of most messaging systems and the prior notification semantics of MMS, and the messages can flow on the MM3 interface.

RTP and MMS

The real time transport protocol (RTP) is an Internet protocol that is not message-based, but is content rich like MIME e-mail. If we think of RTP as carrying live action; then a mapping of the live action onto MMS is analogous to an instant reply snapshot.

RTP carries each medium in a separate stream. Thus, an audiovisual RTP encoding would be two RTP streams: one for the audio and one for the video. The common header for all frames in all RTP streams is given in Figure 7.2.

Multimedia streams are synchronized by means of a timestamp that appears as the header of each packet. The timestamp in each stream is synchronized to a common wall clock time with RTP control packets, and then resulting wall clock times are used to synchronize the streams.

The encoding of particular sources into the payload portion of an RTP frame is covered in source-specific profiles. Of particular interest to MMS application developers is the profile for audiovisual streams: "RTP Profile for Audio and Video Conferences with Minimal Control" (RFC 1890). Profiles for Global System for Mobile Communication (GSM) channel data and GSM audio data have been map defined. See for example, "Real-Time Transport Protocol (RTP) Payload Format and File Storage Format for the Adaptive Multi-Rate (AMR) and Adaptive Multi-Rate Wideband (AMR-WB) Audio Codecs" (RFC 3267).

Figure 7.2 Real-time protocol (RTP) header.

```
 0                   1                   2                   3
 0 1 2 3 4 5 6 7 8 9 0 1 2 3 4 5 6 7 8 9 0 1 2 3 4 5 6 7 8 9 0 1
+-+-+-+-+-+-+-+-+-+-+-+-+-+-+-+-+-+-+-+-+-+-+-+-+-+-+-+-+-+-+-+-+
|V=2|P|X|  CC   |M|     PT      |       sequence number         |
+-+-+-+-+-+-+-+-+-+-+-+-+-+-+-+-+-+-+-+-+-+-+-+-+-+-+-+-+-+-+-+-+
|                           timestamp                           |
+-+-+-+-+-+-+-+-+-+-+-+-+-+-+-+-+-+-+-+-+-+-+-+-+-+-+-+-+-+-+-+-+
|           synchronization source (SSRC) identifier            |
+=+=+=+=+=+=+=+=+=+=+=+=+=+=+=+=+=+=+=+=+=+=+=+=+=+=+=+=+=+=+=+=+
|            contributing source (CSRC) identifiers             |
|                             ....                              |
+-+-+-+-+-+-+-+-+-+-+-+-+-+-+-+-+-+-+-+-+-+-+-+-+-+-+-+-+-+-+-+-+
```

Suppose now that we have two synchronized streams used to display an instant replay from a sports event. One stream is the audio, and the other is the video. Each MMS slide in our instant replay will be a frame from the video stream and a sequence of frames from the audio stream. Figure 7.3 is an example of the data needed to form one MMS slide from a real-time feed.

Video		Snap at 05:06:01	
Audio	Start at 05:05:15		End at 05:07:12

Figure 7.3 MMS snapshot of real-time multimedia stream.

An MMS instant replay might be five, ten, or more individual snapshots, depending on the subject. A soccer goal might be a small number of slides, whereas a breaking news event might be somewhat more slides.

What is really sent to the mobile, as in the MMS message, is a session description protocol (SDP) file that describes the streaming session. The MMS subsystem of the mobile extracts the particulars of the stream from the SDP file and goes back out on the network to start the playback by obtaining the necessary files. Note that this is similar to how we got from the MMS notification message to the MMS itself. There is no problem in computing that cannot be solved with an additional level of pointers.

A rather critical issue that we duck in this technical description is the origin of the snapshot descriptions. One imagines a new kind of media editing job, much like the folks that already edit instant replays in sports telecasts and news clips for news programs. Somewhere, this job is already coming into being.

SMIL Templates and Automatic MMS Production

Whether the legacy data are coming from an e-mail server or from a real-time data feed, a critical part of moving the data across the MM3 interface to the MMSC and on to the recipient is encapsulating the data in an MMS envelope. This means both providing values for the various MMS header fields and composing a synchronized multimedia integration language (SMIL) script that organizes the various components of the data. The filling in of the header fields is a straightforward task of mapping message meta-information from one protocol to another: address-to-address, sender-to-sender, subject-to-subject, and the like. The challenge is the SMIL.

Clearly, this will not be done by hand. Rather, a formula of some sort will be applied to the incoming data that results in the SMIL script that wraps it. One form of such a formula is a parameterized SMIL template. The arriving data supply values for the parameters and out they go, wrapped in their own cozy SMIL script.

The following is an example of such a SMIL template written in Perl. The template takes an audio track, the duration of the audio track in milliseconds, and three image file names as its arguments, and outputs the SMIL program that shows the three pictures, one after another for an equal amount of time, and plays the sound track underneath. The output of this program could be fed directly to the SMIL2MIME program from Chapter 3 to create the outgoing MIME file.

```
my ($sound_track, $sound_track_length, $picture1, $picture2,
    $picture3) = @ARGV;

my $picture_length = $sound_track_length/3;

my $last_picture_length = $sound_track_length-2 *
   $picture_length;

print << "EOF";
<smil>
    <head>
        <layout>
            <root-layout/>
                <region id=\"full_screen\" top=\"0\" left=\"0\"
                        height=\"100%\" width=\"100%\"/>
        </layout>
    </head>
    <body>
        <par dur=\"$sound_track_length ms\">
            <seq>
                <par dur=\"$picture_length ms\">
                    <img src=\"$picture1\"
```

```
                    region=\"full_screen\"/>
                </par>
                <par dur=\"$picture_length ms\">
                    <img src=\"$picture2\"
                     region=\"full_screen\"/>
                </par>
                <par dur=\"$last_picture_length ms\">
                    <img src=\"$picture3\"
                     region=\"full_screen\"/>
                </par>
            </seq>
            <audio src=\"$sound_track\"/>
        </par>
    </body>
</smil>
EOF
```

Figure 7.4 illustrates how this template might be used.

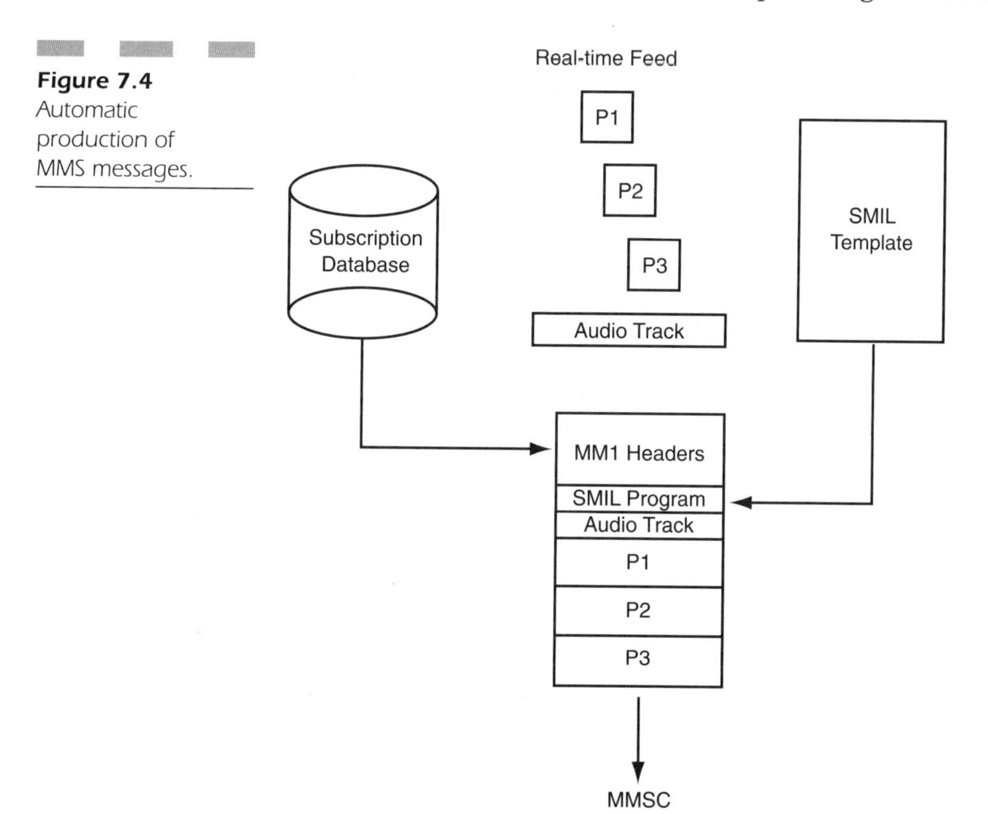

Figure 7.4 Automatic production of MMS messages.

Boston Bruins Example

This type of MMS template could be applied to any number of sports and broadcast events to transmit highlights to the mobile phone as the following hypothetical scenario illustrates:

A real-time feed of a goal during a Boston Bruins hockey game comes in on RTP. A number of people have signed up on the Boston Bruins home page (http://www.bostonbruins.com) to receive MMS messages whenever the Bruins score a goal. They pay the Bruins $10 per month for this service. The Boston Bruins owners take a $1 cut and pass $9 onto MMS Sports, the company that actually sends out the message. Of course, MMS Sports will have to provide editing services and pay the operator for sending the MMS message.

A program such that described in the previous section combines the three pictures and the real-time audio track of the Bruins' radio broadcast of the goal to produce the MIME package. Every time a goal is scored, the subscriber database is scanned for everyone who has paid to receive an MMS of all Bruins' goals. Each subscription is turned into a set of MMS header fields and sent to Boston's T-Mobile MMSC. Within less than a minute of the goal, Bruins' fans all over the world are seeing the instant replay stills on their mobile phones and hearing the radio announcer's golden tones describe the play.

As we will see later, the enterprise running this service is probably connected through an MM7 interface to the T-Mobile MMSC and, therefore, uses MM7 header fields. For this chapter, it is important to understand that the only practical way to support such a subscription service is to rely on MMS messages that are machine generated using a pre-prepared SMIL template.

It is clear that the procedure described in this chapter can be generalized and improved. We could handle any number of pictures and also probably get the duration of the audio track out of the audio file itself. We could also drive this whole process through a Web page that lets the user pick or upload a number of images, add some text, and maybe even send along an audio track, to compose a structured MMS message that would be sent to a mobile phone number he entered. A number of network operators are already offering a service like this on their Web sites.

RTP Example Summary

Mapping one messaging protocol to another, such as the previous example of mapping between e-mail and MMS, is an obvious and straightforward use of the MM3 interface. A connection between MMS and another messaging system really has to do little more than reformat message parts and change the addressing scheme.

The RTP example is intended to demonstrate that you can map protocols that are not intrinsically messaging protocols into MMS too. It requires a little more creativity both on the technical side and on the business side, but where there are challenges, there are opportunities. In fact, there is a good chance that you thought of a couple of better ways of mapping real-time feeds onto MMS while you were reading the RTP-to-MMS example.

The RTP example also shows that MM3 is not necessarily a two-way connection like MM1 and MM7. It makes a certain amount of sense to frame-grab a video stream and package the frames up with the audio and send the result out as an MMS message. It makes somewhat less sense to convert an MMS message into an RTP stream.

Databases and MMS

On a much more mundane, but perhaps more practical plane, databases are classic legacy IT resources that might contain content that would be useful to deliver to a mobile device.

Imagine, for example, sending an SMS to your home database of digital photographs and calling forth on your mobile phone pictures from little Bobby's recent birthday party. This sure beats carting around a photo album or a wallet full of dog-eared snapshots. Or suppose you wanted to relight the pilot light on your hot water heater and had lost the instructions. You could send an SMS containing the model number to the telephone number on the heater and get back an MMS that contained pictures with audio instructions on how to perform the delicate task, without launching yourself into the next county.

The MM3 impedance matching being done here is between structured query language (SQL) and MMS. There are many ways of doing this. For example, if the SQL query yielded only one result, then each MMS frame could be one value in the record—first frame is NAME: Sally Green, next frame is AGE: 32, and so forth. Alternatively, if the query returned multiple records, each MMS frame could be one record.

MMS Message Store

In the previous two examples, we created the MMS message "on the fly" from the incoming data. In the first case, it was e-mail. In the second case, it was a real-time video feed. The translation from one protocol to another was the service performed across the MM3 interface. The translation was done "on the fly" because the MMS message was, in a sense, a continuation of the communication. The communication was forwarded onto the mobile network in the form of an MMS message.

The MM3 interface supports another kind of commercial MMS application, in which the translations to MMS format are done ahead of time, typically with much care, and then sent out on an as-needed basis. In this case, MM3 does not connect two protocols. Instead, it connects MMS to the message store. It is much more like a database connection than a communication protocol connection.

If you plan to be in the MMS content business, as opposed to the MMS service business, somewhere in your architecture you will have an MMS message store. Figure 7.5 is a simple diagram of an MMS content service. The MMS message store is a holding tank for MMS messages between the time they are created and sent to an end user. The content of the message store is not transient, as in the case of e-mail or a replay, but somewhat more permanent, such as a newspaper's archive of news photographs or a police archive of mug shots. You can probably think of better examples and maybe you have in mind starting a content service of some sort.

An MMS message store can hold MMS messages in one of three basic formats:

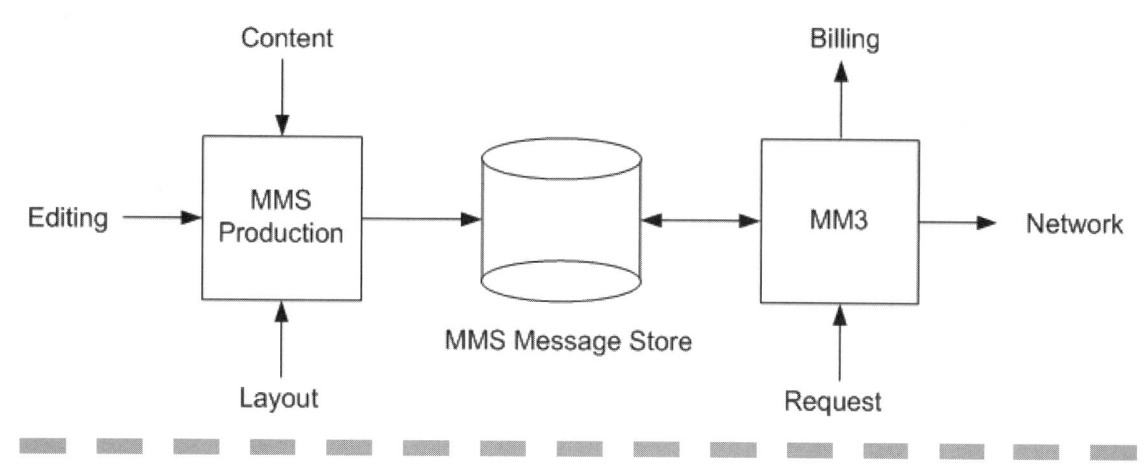

Figure 7.5 MMS production process.

1. Individual media components.
2. MIME-encoded MMS messages.
3. MMS encapsulation protocol-encoded MMS messages.

Which format you use for your content service depends on the nature of the content itself and what kind of flexibility you want to provide in distributing it.

At one extreme, the MMS message store contains only a large collection of images, audio tracks, and text blurbs. An incoming request gets translated into a SMIL script that collects its constituent elements from the message store as it gets turned into an MMS message. This approach offers maximum flexibility, but as with all flexible solutions, shifts the burden of defining exactly what is in the MMS message from the content provider to the user. It also requires the greatest amount of computing per MMS message because each one has to be assembled "on the fly."

A more reasonable approach is to have the SMIL scripts prebuilt, indexed, and classified somehow. A request to the service is then phrased in terms of this indexing, rather than in terms of the media components of the response. A key value-add of the content service, in this case over and above the media elements, is the indexing and the layout of the response.

Whether you go ahead and build the MMS message that is described by each SMIL script depends on the relationship of media elements to scripts. If each media element appears in only one script, you may as well build the messages. If, on the other hand, media elements typically appear in many scripts, you would be making multiple copies of the element if you created all the MMS messages. Furthermore, if you decide to update a particular element—you find a better picture of something, for example—you have to find all the MMS messages that contain the element and rebuild them with the updated element.

If you decide to build the messages, you can store them either in the MIME format or MMS encapsulation protocol, encoded and ready to go. Which of these two alternatives you pick will be determined primarily by the technical and business properties of your network interface. Some network interfaces only accept the MIME format and insist on doing the MMS encapsulation protocol encoding themselves. Other interfaces are quite content with accepting the MMS encapsulation protocol encoding.

In this latter category are wireless access protocol (WAP) gateways. If you send out the MMS notification message yourself, with a location URL back to your MMS message store, you may want to have the MMS message itself ready to go when the WAP gateway comes for it.

The MMS architecture and the MM3 interface handle any one of these MMS message store strategies handily. The location URL in an MMS notification message is typically a direct file reference, but it does not have to be. It can, for example, be a query that describes the MMS that is to be returned. Based on this query, your MMS message store can simply send back a ready-to-go WSP-encoded MMS message, or it can scramble and find the SMIL script and all its parts, and assemble the MMS message "on the fly."

Summary

MM3 is a rather abstract and loosely defined interface for connecting the MMS messaging architecture to other information handling architectures. In fact, it is really little more than a reference point on the MMS architectural diagram because each matching between MMS and another protocol will have to take into account particular properties of both protocols.

In a way, MM3 is an impedance matcher or protocol translator between the world of mobile multimedia services and the rest of the world. The MM3 interface can be used by all types of content providers wishing to extend their service offering into the mobile domain using the MMS.

CHAPTER 8

The Value-Added Service Provider Interface—The MM7 Interface

MM7 is the future. MM7 is not designed to handle legacy content like the messages that come in over the MM3 interface. It is the interface created to be multimedia messaging service (MMS) from the beginning. MM7 supports commercial and high-volume throughput, unlike the one-at-a-time, look-what-I-can-do-with-my-phone messages that come in over the MM1 interface.

MM7 is where the portals and enterprises will connect, and where the operators themselves will offer MMS-creation services to their customers. MMS sending and receiving entities on the MM7 interface are called value-added service providers (VASPs).

The actual MMS messages traveling through MM7 are the same as the messages traveling over MM1 and MM3. They are MIME-encoded, SMIL-synchronized messages described in detail in Chapter 3. And when they go out on the air interface, they are MMS encapsulation protocol-encoded like the others too; after all, the protocol used by the multimedia messaging service center (MMSC) to deliver the message to the receiving handset is MM1.

What is different about MM7 is the amount of information in the envelope that is carrying the message. This information is a conversation between the operator's MMSC and the content provider's server. And on a commercial interface such as MM7, they have a lot to say to each other. Issues of billing, authentication, administration, and transaction management must be, and are, addressed on the MM7 interface. The MM7 conversation is more of a wholesale, business-to-business, peer-to-peer conversation compared to the MM1 interface, which is a retail, business-to-customer conversation.

Because of the difference in the standing of the two parties across MM1 and MM7 interfaces, things can be accomplished on MM7 that are impossible on MM1. One primary example of such a difference is transcoding. There is no way on the MM1 interface for a customer to instruct the network operator to refrain from transcoding a message; that is, to ensure that the MMS message reaches the recipient as submitted. Providing an indication that an MMS message "should not be converted or changed by the MMS service provider before it is delivered to the recipient" is explicitly provided for on the MM7 interface.

The difference between the two interfaces extends all the way to the recipient of the MMS. As you may recall, the recipient of an MMS from the MM1 interface can indicate that no delivery or read report is to be returned to the sender, even if the sender has asked for it. If the MMS message has come from the MM7 interface, however, this indication will be ignored by the network operator and a delivery report will be returned to the VASP, whether the recipient likes it or not. In the current implementation of MM1, a handset can block a read receipt, as it is sent as a separate MMS message, and the recipient would be charged for sending that MMS. That will change as implementations catch up to the latest specifications, but it will take time.

As a result, the MM7 conversation needs richer syntax and semantics than just hypertext transfer protocol (HTTP) header fields with values to carry on this conversation. And it needs to be full duplex in the sense that either partner may start a conversation. Finally, it will have conversational elements that do not include an outgoing or incoming MMS message, but rather have to do with the business between the two entities.

MM7 uses a header language called the simple object access protocol (SOAP). SOAP is an extensible markup language (XML) protocol that is transported over HTTP. In essence, we are putting the MMS inside another envelope that is then carried by HTTP. This is the SOAP envelope. Figure 8.1 is the envelope diagram for the MM7 interface.

You may still find some proprietary MM7 implementations in the field that are based on the simple mail transfer protocol (SMTP). They are not covered here because they are not standards-compliant, and the SOAP version of MM7 is where everyone is heading, even if they are not there yet.

Besides just wanting to say more to each other, the VASP and operator want to maintain the context of an ongoing conversation. For example, the MMSC might want to say back to the VASP, "Remember the MMS you sent out yesterday to 43 people? Well, two more of them just picked up the MMS you sent." The SOAP schema for MM7 includes the notion of a session, including a session identifier, to do this.

Submit on the MM7 Interface

The submission of an MMS by a VASP to the MMSC using the MM7 interface and SOAP is shown in the following code. This example

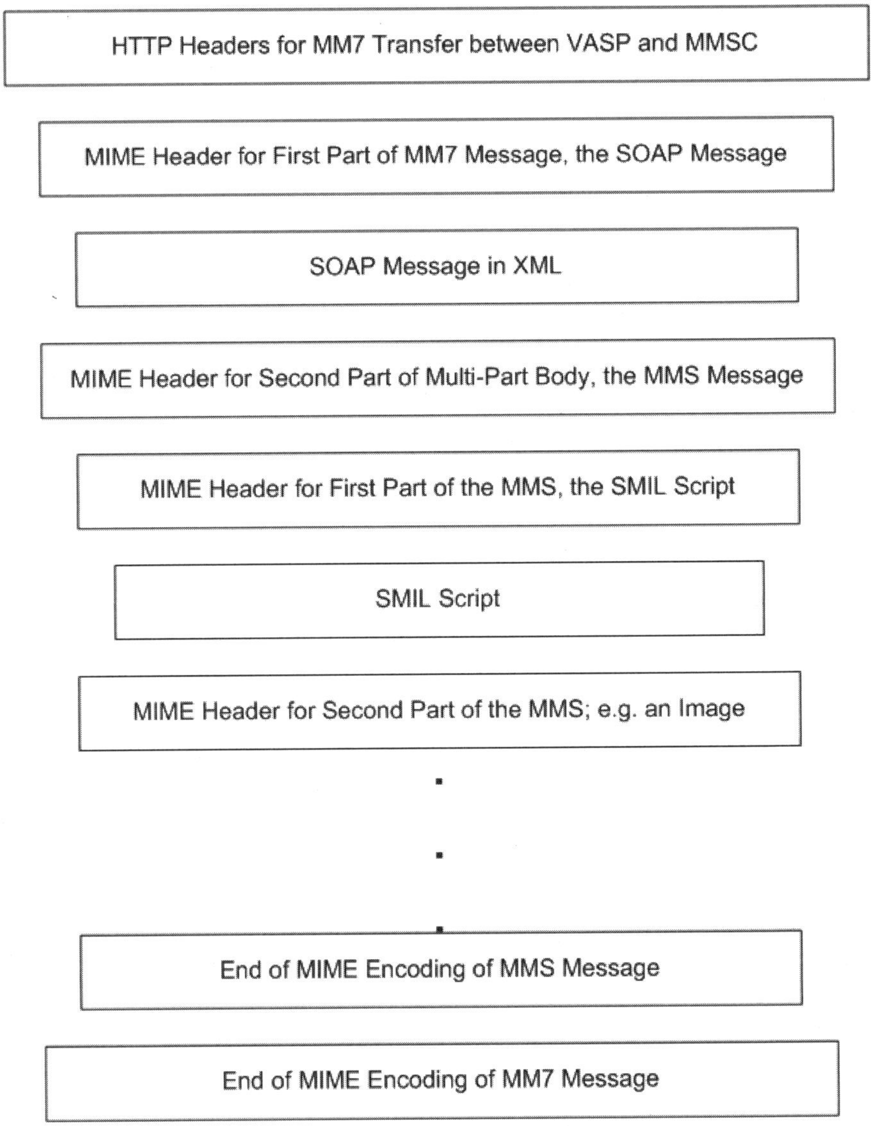

Figure 8.1 Envelopes on the MM7 interface.

HTTP Headers for MM7 Transfer between VASP and MMSC

MIME Header for First Part of MM7 Message, the SOAP Message

SOAP Message in XML

MIME Header for Second Part of Multi-Part Body, the MMS Message

MIME Header for First Part of the MMS, the SMIL Script

SMIL Script

MIME Header for Second Part of the MMS; e.g. an Image

.

.

.

End of MIME Encoding of MMS Message

End of MIME Encoding of MM7 Message

appears in European Telecommunications Standards Institute (ETSI TS) 23.140.

This message is being sent from the VASP to the MMSC using HTTP, so the very outermost structure is a set of HTTP message headers, followed by a blank line and then the HTTP message body. The HTTP headers, among other things, say that the body consists of a number of parts (as it turns out, exactly two), and that these parts are separated by the MIME boundary separator "NextPart_000_0028_01C19839.84698430." Note that in this example, the MMS message does not contain a SMIL presentation element, only the media content. Therefore, the MIME type is expressed as "multipart/mixed" instead of "multipart/related."

The two parts of the message body are: 1) the VASP's instructions to the network operator rendered in the SOAP language regarding how to handle the MMS, and 2) the MMS message itself. Recall that the MMS is in turn a collection of parts separated by MIME boundaries. In the example, the boundary separator of the parts of the MMS is "StoryParts 74526 8432 2002-77645."

We know all about the second part, the MMS message, from previous chapters. Therefore, we will concentrate on the first part, which is the MM7 conversation between the VASP and the MMSC.

We still have to say most things with plain old header fields on the MM1 and MM3 interfaces such as From, To, and Subject. But on MM7, we say them in SOAP, rather than as HTTP field/value pairs:

```
POST /mms-rs/mm7 HTTP/1.1
Host: mms.omms.com
Content-Type: multipart/related;
   boundary="NextPart_000_0028_01C19839.84698430"; type=text/xml;
     start="</tnn-200102/mm7-submit>"
Content-Length: nnnn
SOAPAction: ""

--NextPart_000_0028_01C19839.84698430
Content-Type:text/xml; charset="utf-8"
Content-ID: </tnn-200102/mm7-submit>

<?xml version='1.0' ?>
<env:Envelope xmlns:env="http://schemas.xmlsoap.org/soap/
       envelope/">
      <env:Header>
          <mm7:TransactionID
xmlns:mm7="http://www.3gpp.org/ftp/Specs/archive/23_series/23.140
           /schema/REL-5-MM7-1-0" env:mustUnderstand="1">
                vas00001-sub
          </mm7:TransactionID>
```

```
        </env:Header>
        <env:Body>
            <mm7:SubmitReq
xmlns:mm7="http://www.3gpp.org/ftp/Specs/archive/23_series/23.140
            /schema/REL-5-MM7-1-0">
                <MM7Version>5.3.0</MM7Version>
                <SenderIdentification>
                    <VASPID>TNN</VASPID>
                    <VASID>News</VASID>
                </SenderIdentification>
                <Recipients>
                    <To>
                        <Number>7255441234</Number>
                        <RFC2822Address
                         displayOnly="True">7255442222@OMMS.com</R
                         FC2822Address>
                    </To>
                    <Cc>
                        <Number>7255443333</Number>
                    </Cc>
                    <Bcc>
                        <RFC2822Address>7255444444@OMMS.com</
                         RFC2822Address>
                    </Bcc>
                </Recipients>
                <ServiceCode>gold-sp33-im42</ServiceCode>
                <LinkedID>mms00016666</LinkedID>
                <MessageClass>Informational</MessageClass>
                <TimeStamp>2002-01-02T09:30:47-05:00</Date>
                <EarliestDeliveryTime>2002-01-02T09:30:47-
                 05:00</EarliestDeliveryTime>
                <ExpiryDate>P90D</ExpiryDate>
                <DeliveryReport>True</DeliveryReport>
                <Priority>Normal</Priority>
                <ChargedParty>Sender</ChargedParty>
                <DistributionIndicator>True</
                 DistributionIndicator>
                <Subject>News for today</Subject>
                <Content href="cid:SaturnPics-
                 01020930@news.tnn.com"; allowAdaptations="True"/>
            </mm7:SubmitReq>
```

```
    </env:Body>
</env:Envelope>

--NextPart_000_0028_01C19839.84698430
Content-Type: multipart/mixed; boundary="StoryParts 74526 8432
      2002-77645"
Content-ID:< SaturnPics-01020930@news.tnn.com>

--StoryParts 74526 8432 2002-77645
Content-Type: text/plain; charset="us-ascii"

Science news, new Saturn pictures...

--StoryParts 74526 8432 2002-77645
Content-Type: image/gif;
Content-ID:<saturn.gif>

R0lGODdhZAAwAOMAAAAAAIGJjGltcDE0OOfWo6Ochbi1n1pmcbGojpKbnP/
      lpW54fBMTE1RYXEFO
...

-StoryParts 74526 8432 2002-77645--
--NextPart_000_0028_01C19839.84698430--
```

The HTTP response sent back from the network operator to the VASP looks like this:

```
HTTP/1.1  200 OK
Content-Type: text/xml; charset="utf-8"
Content-Length: nnnn

<?xml  version='1.0' ?>
<env:Envelope xmlns:env="http://schemas.xmlsoap.org/soap/
       envelope/">
      <env:Header>
         <mm7:TransactionID
xmlns:mm7="http://www.3gpp.org/ftp/Specs/archive/23_series/23.140
      /schema/REL-5-MM7-1-0" env:mustUnderstand="1">
               vas00001-sub
         </mm7:TransactionID>
     </env:Header>
```

```
    <env:Body>
        <MM7Version>5.3.0</MM7Version>
        <Status>
            <StatusCode>1000</StatusCode>
            <StatusText> Success</StatusText>
        </Status>
        <MessageID>041502073667</MessageID>
    </env:Body>
</env:Envelope>
```

In a nutshell, SOAP is a full-bodied XML dialect, rather than the sequence of name/value pairs found on the MM1 and MM3 interfaces. The core definition of SOAP used on MM7 is SOAP 1.1 defined in: W3C Note 08 May 2000, "Simple Object Access Protocol (SOAP) 1.1," URL: http://www.w3.org/TR/SOAP.

The dialect of SOAP used on the MM7 interface has its own schema: http://www.3gpp.org/ftp/Specs/archive/23_series/23.140/schema/REL-5-MM7-1-0.

MM7 Messages

In spite of the fact that the syntax and semantics of the messages on the MM7 are more complex than the messages on the MM1 or MM3 interfaces, there really are not that many more messages than on the MM1 interface, and many of them serve the same purpose. Table 8.1 lists all the messages on the MM7 interface, as of Release 6 of the MMS specifications.

TABLE 8.1 Messages on the MM7 Interface

MM7 Message
Submit
Delivery
Cancel
Replace
Delivery Report
Read Reply Report

Of particular interest to MMS application developers is the Submit request because this envelope carries the content. Table 8.2 lists the information that a VASP can provide to the network operator when sending an MMS, along with comments on elements that are unique to the MM7 interface.

TABLE 8.2
Submit Fields Unique to MM7

Information Element	Description (if different from MM1)
Transaction ID	
Message Type	
Interface Version	
VASP ID	Unique identifier of the VASP.
VAS ID	Unique identifier of the VASP's application.
Sender Address	
Recipient Address	
Service Code	Charging code provided by the VASP.
Linked ID	Links this submission with a previous submission.
Message Class	
Date and Time	
Time of Expiry	
Earliest Delivery Time	
Delivery Report	
Read Reply	
Reply-Charging	
Reply-Deadline	
Reply-Charging-Size	
Priority	
Subject	
Adaptations	If set to "FALSE" no transcoding is to be done to the message.

(continued on next page)

TABLE 8.2
Submit Fields Unique to MM7 (continued)

Information Element	Description (if different from MM1)
Charged Party	Indication of which party—sender, recipient or both—is to be charged for the MMS.
Content Type	
Content	
Message Distribution Indicator	If set to "FALSE," content is not to be redistributed.

As you might imagine, many of these fields are little more than toeholds for functionality that is expected to emerge. Indeed, one could argue that the use of SOAP on the MM7 interface was more a matter of being future-proof than a necessity for today's capabilities.

The Adaptations field has already been described. In many cases, the VASP will know exactly what handset the recipient is carrying, and just as importantly, how it is configured. The VASP has every incentive to carefully tailor the message to the recipient's equipment for maximum pleasure and efficiency. VASPs do not want any off-the-shelf transcoding engine between themselves and the recipients to undo their careful work.

The VASP ID, VAS ID, Service Code, and Charged Party fields are the harbingers of the business relationship between the VASP and the network operator. It is fully expected that such relationships will be very complex and will be driven by all sorts of parameters of the MMS messages. Some of these parameters, such as the VAS ID and the Service Code, are set by the VASP, either for its own records or to inform the terms of the business agreement.

The Linked ID is a number provided by the MMSC side of the MM7 interface as a way to link the incoming submission being provided by the VASP to a previous message from the MMSC to the VASP. Unlike the MM1 interface, which is stateless, the MM7 interface supports the concept of a persistent context for a series of message exchanges between the VASP and the MMSC. Since the VASP and the MMSC may have a number of these going at any one time, each of them must be identified. The Linked ID is the identifier of a messaging context between the MMSC and the VASP. It should be noted that currently, only the MMSC, not the VASP, can initiate such a session identifier on the MM7 interface.

One scenario that calls for the creation of a communication context is when a user sends an MMS message to the VASP requesting some service from one of the VASP's applications. In this case, the MMSC will generate a Link ID and include it in the deliver message that it uses to pass the user's request to the VASP. The VASP will then put that Link ID into the response to the end user. This lets the network operator associate an outgoing VASP message with a specific user request, and thereby puts the operator in a position to block VASP messages that are not associated with a user request; in other words, what is from the operator's point of view an unsolicited message or SPAM.

The Link ID is really a SPAM control. It is a ticket that the MMSC gives to a VASP that lets him send a message. VASPs may initially be able to send unsolicited messages, but if they abuse this privilege, they will only be able to send messages with Link ID tickets.

Copyright Protection and Digital Rights Management Issues

The final field that is new in the MM7 submit message is the Message Distribution Indicator. While much concern is expressed about the protection of the digital rights surrounding the content in MMS messages, no one wants to accept primary responsibility for the protection of these rights, or more properly, the liability for their violation.

The Message Distribution Indicator is an acknowledgment that something needs to be done in the area of digital rights management, but it is little else. A VASP can set this bit on the envelope transmitting an MMS to the MMSC. And the MMSC, on behalf of the network operator, can use it, for example, to deny requests from customers to forward the MMS to their e-mail account, rather than downloading it to their phone.

Once they have downloaded the MMS to their phone, it becomes more complicated. Should they now be able to forward the MMS to their e-mail account? If not, then either their handset will run out of memory rather quickly, or all the content they are purchasing is ephemeral.

Currently, all of this is only grist for discussion because the Message Distribution Indicator bit does not go through the MM4 interface between MMSCs, nor does it ever get to the handset over the MM1 interface. Furthermore, the current MM1 interface, as defined by the Open Mobile Alliance (OMA) MMS specifications, does not support the ability to forward an MMS without actually retrieving the message. This should change in future versions of the OMA MMS specifications.

At any rate, copyright protection and digital rights management are clearly more complicated than one bit. The Message Distribution Indicator is a "thumb in the dyke" until we can figure out what to do.

Security

MM7 is a high-volume interface and a privileged interface to the MMSC. As a result, it is fully expected to also be a secure interface. Since it would not be economical to build special communication channels between VASPs and the MMSCs, it is also expected that technically, MM7 will be implemented on top of Internet connections. The use of HTTP and SOAP probably showed you this.

HTTP, as well as the underlying Internet protocols, offer a number of secure channel capabilities that include mutual authentication, message integrity, and confidentiality. Which of these are used and what the procedures for using them are will vary from network operator to network operator. The MM7 specifications explicitly leave the security of the communication channel between them up to the communicating parties.

Summary

This chapter covered the high-speed, high-volume, business-to-business interface between the MMS system and the VASPs. Many people think this interface, and not the MM1 person-to-person interface, will determine the success or failure of MMS.

MMS is about content, and with all due respect to little Bobby and his friends, content is more than pictures of little Bobby's birthday party. Content is about selection, organization, presentation, refinement, and branding—all the things that go into a movie, a recording, or a book that are missing from family snapshots.

There are certainly those who think MMS is just short messaging service (SMS) with pictures. If this is true, or if operators and content providers cannot find a mutually beneficial business model, then the MM7 interface will gather dust.

If MMS application developers discover ways of presenting existing content on this new channel, or even better yet, discover new content that was just waiting for this channel to exist, then MM7 will be to the mobile telephone system what a broadcast license is to the television industry.

CHAPTER 9

The MMS Business Case

The technical infrastructure and standards for multimedia messaging service interoperability may still be in flux, but that has not deterred the analysts from making revenue projections reaching into the billions of dollars. According to Juniper Research, MMS has the potential to generate $8.3 billion for wireless carriers and their partners during 2004. Even the more conservative estimates agree that after a relatively slow ramp up in 2003, MMS application traffic will begin to equal and will ultimately overtake the phenomenal success of short messaging service worldwide. Frost & Sullivan predicts that MMS will account for two thirds of all mobile messaging revenue (excluding e-mail) by 2006.

Wireless carriers, still smarting from the failure of overhyped wireless application protocol (WAP) applications a few years ago, are reluctant to discuss specific internal revenue goals, or even the rate of MMS adoption by their customers. Nevertheless, carriers worldwide are investing significant resources in rolling out MMS services, with over 60 carriers announcing MMS launches in the course of 2002, and even more ready to hit the market in 2003.

The carriers are not the only group with a serious stake in the success of MMS. Mobile device makers are counting on consumer desire to access multimedia applications and snap pictures with their phones to drive consumer purchasing of high-end MMS-capable, camera-integrated handsets. Network infrastructure providers, such as multimedia messaging service centers, are waiting for the predicted surge in higher priced MMS traffic to recoup their investments in upgraded gateways, and to convince the remaining wireless carriers to commit to MMS infrastructure services. Online, broadcast, and print content providers are looking for ways to extend their brand into the wireless world via MMS. Aggregators and portals are crafting new business projections. And mobile application developers are looking for the best place to enter the MMS value chain. This chapter analyzes the relationships among the major MMS stakeholders, reviews current pricing and application distribution trends, and summarizes the business decisions and partnership opportunities facing MMS application developers.

The MMS Value Chain

The wireless carriers and mobile device makers have a clear stake in the revenues generated by MMS adoption, as well as in the infrastructure required to deliver MMS messages to customers. But owning an MMS-capable handset and being a customer of an MMS-providing carrier are not in themselves sufficient to motivate end users to try and buy MMS services. In fact, the Juniper Research estimate cited previously assumes that the bulk of revenue will be generated by MMS services based on commercial content, rather than by customers sending messages to each other. Of the $8.3 billion total MMS revenue Juniper forecasts for 2004, peer-to-peer messaging revenues will amount to $2.7 billion, while server-to-mobile revenues will account for the remaining $5.6 billion. Even if you take these billion dollar estimates with a grain of salt, it seems clear that content servers, content owners, and application developers will drive considerably more MMS revenue than swapping of photos and informal MMS messages among subscribers. Therefore, carriers and device manufacturers have to work with a number of business and technical partners. These partners are the components of an emerging MMS value chain.

Convincing tens of millions of consumers to download hundreds of millions (and ultimately billions) of MMS messages will require content and applications that deliver on the promise of wireless multimedia. In this situation, content providers and application developers are just as necessary to MMS success as are the carriers and the handset makers. Mass adoption will also require interoperability among MMS devices, roaming agreements so that customers can easily exchange MMS messages and applications with each other, and ways to store and retrieve favorite MMS sessions that will not all fit on the user's phone. Aggregating a variety of applications and MMS content will require testing and certification that third-party developer applications will work with different carrier and handset application program interfaces (APIs). From a marketing and sales perspective, it will be desirable to have MMS portals on the Web to serve as entry points for occasional or potential users who do not yet own an MMS phones, and who want to test the service or send a message to an MMS-enabled friend.

In fact, as Figure 9.1 illustrates, the MMS value chain includes new service providers who are still striving to define their piece of the MMS action, as well as fundamental components such as wireless carriers and device makers. This jostling for a place on the value chain is typical at the roll out of any major new technology, and it is likely that many of the would-be middlemen offering slices of MMS service will consolidate or disappear in the next few years. Rather than covering each value chain opportunity in detail, this chapter provides a strategic analysis of the players and business opportunities of vital interest to MMS application developers.

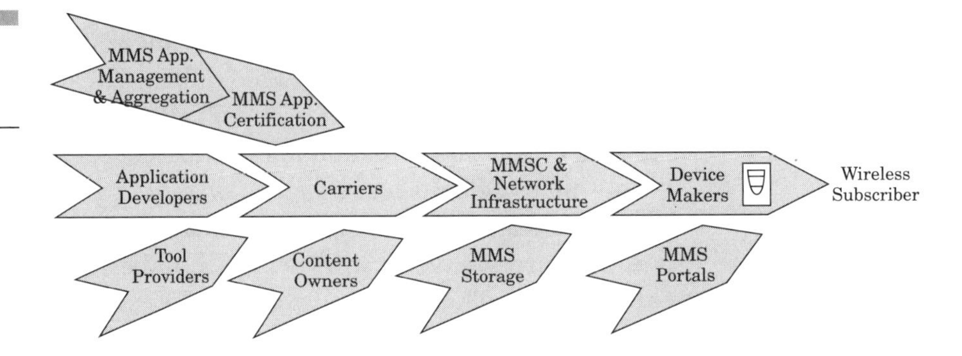

Figure 9.1 MMS value chain participants.

Strategic View of Value Chain Participants

The key to analyzing MMS business models is to understand what specific functions each participant contributes to the MMS value chain, and how that participant expects to turn its contribution into revenue. In some cases, such as the wireless carrier, the functions are obvious. In others, such as an MMS portal or storage service, the long-term need and value of the function are still in question. Table 9.1 provides an overview of the functions as presently defined, along with a summary of opportunities, threats, and revenue options for each value chain component.

TABLE 9.1 Value Chain Analysis

Component	Major Functions	Opportunities	Threats	Revenue Options
Wireless Carriers	Air interface to carry MMS.	New services based on MMS traffic and applications.	Low consumer adoption of MMS.	Fee per MMS; premium content and subscription fees; roaming and MMS storage services.
Device Makers	Design and manufacture of MMS devices.	Mass upgrade to MMS devices boosts sales; higher end devices expand traditional phone functionality and market (cameras, music, etc.).	Consumer reluctance to pay for higher priced handsets leads to overstocks and discounting.	Sale of devices; becoming a channel for application distribution (Club Nokia, Club Samsung, etc.).
MMS Infrastructure Providers	MMSC connection between carriers and devices, legacy interfaces, traffic monitoring, and accounting.	Establish a place in all MMS distribution and delivery systems.	Competition among different providers and replication of nodes bogs down delivery systems.	Direct sale of hardware and software to carriers; fee per message based on traffic handled centrally.
Application Management and Certification Providers	Testing and validating application performance; app database systems and tools.	High demand for applications creates a seller's market for app-related services.	Slow MMS adoption reduces demand for middleman services.	Fees for testing, certifying, and handling applications.
Content and Application Aggregators	Agreements with app and content providers; establishing APIs.	Ability to scale with demand for applications.	Competition drives down prices and fees.	Percentage of application revenues based on traffic or licensing to carriers and/or advertising sponsors.
Content Owners	Providing legacy or new content.	New outlets for existing content.	Loss of control of content.	Licensing and royalty fees.
Application Developers	Providing MMS apps.	New field for application sales.	Glut of apps or lack of adoption drives down developer revenues.	Licensing or revenue sharing agreements with multiple service providers.
Tool Providers	Providing development and presentation tools for MMS apps and content.	OEM agreements with handset makers; sales and licensing to developers and infrastructure providers.	Proprietary MMS tools are built into handsets and branded development suites (Adobe, Microsoft, etc.).	OEM, licensing and sales of tools.

(continued on next page)

TABLE 9.1 Value Chain Analysis (continued)

Component	Major Functions	Opportunities	Threats	Revenue Options
MMS Storage Services	Temporary or archival storage for MMS files.	Size of MMS messages drives demand for new types of image storage services.	Storage needs met by carriers or other service providers.	Fee per message stored and retrieved, or monthly storage fees.
MMS Portal Services	Access to MMS services (sending messages, accessing stored messages) via the Web.	Demand for exchange of MMS among users who do not own MMS-capable phones.	Cost of sending MMS limits demand, and portal traffic remains minimal.	Fee per message sent or advertising support for selected MMS services.

Pricing Fundamentals

One aspect of the MMS value chain that has yet to be resolved to the satisfaction of all participants is how and when consumers should pay for sending and receiving MMS messages. If the price point is set too high, there is a danger that consumers will drag their feet in adopting MMS services, or worse yet, turn their back on the whole concept of multimedia phones. MMS is, after all, a totally optional service that has yet to develop a mass market following. On the other hand, if the price point is too low, even mass adoption will not generate enough revenues to justify the investment in the MMS infrastructure and resources provided by all the value chain participants. And if content providers and application developers, in particular, do not see a return for their investment, these critical "pump primers" are likely to abandon MMS.

Wireless carriers bear the primary responsibility for balancing consumer price sensitivity with value chain revenue sharing requirements, and for establishing the optimal price point for early MMS services. In theory, carriers have a variety of pricing options for MMS services. They can charge customers according to the size of the MMS message either per message (e.g., $1 for a 100 kB message), or as part of a monthly wireless data allowance (e.g., $50 for 1 MB of data sent and received). Or, they can bundle a set number of MMS messages of any size into a fixed monthly fee (e.g., 50 MMSs for $25). Alternatively, they can package certain MMS services into premium subscriptions with messages included in the subscription fee (e.g., photo swapping with friends and family for $20 per month). Or, simplest of all from the consumers' point of view, carriers can set a fixed price per MMS message sent (e.g., $.0.50 per message) and itemize these messages on the monthly bill.

In practice, many carriers dodged the question of pricing as they introduced MMS services by subsidizing all MMS traffic for subscribers for a specified period. These promotions helped to drive the sales of MMS-capable handsets, which were sometimes also subsidized, and motivated large numbers of early adopters to try MMS. As a marketing device, the free trials were a big success. More than half a million MMS messages were exchanged during the first three months of free access by TIM in Italy, while the rollout of free MMS by Singtel in Singapore stimulated over 250,000 messages. Carriers also used these promotional periods to test the attitudes of early adopters toward proposed pricing schemes, and to gather comparative data from other carriers. The carriers who launched MMS early in 2002—for example Telenor in Norway—released a promotional brochure with a relatively high price for the service (1.30 Euros per message) while it was still free, then reduced the delivery cost to 0.65 Euros by the time it was actually billing customers.

Carriers, in fact, have watched each other's pricing schemes fairly closely. With more than 60 carriers offering MMS services by the end of 2002, the price ranges and variety of pricing schemes look remarkably similar. According to an Ovum research report in December 2002, all European carriers have adopted a fixed cost per MMS message pricing model. Within that model, the majority of carriers are charging between 0.55 and 0.65 Euros per message. The lowest priced European MMS services are in Eastern Europe, with carriers in the Czech Republic charging 0.30 Euros, and the highest price is in Switzerland, where SwissCom is charging 0.80 Euros per message.[1]

Even if the fixed price per message takes hold as a worldwide MMS pricing model, the carriers will be able to experiment with other revenue options. A number of carriers already combine the fee-

[1] John Delaney, "MMS Picture Messaging: Is the Price Right?" Ovum, December 18, 2002 at www.ovum.com/go/ovumcomments/016489.htm.

per-message model with extra charges for premium content. SingTel, for example, charges $3.50 ($ Singapore) for downloading Hallmark branded MMS greeting cards, and in the Philippines, Smart Telecom charges 15 pesos (Philippine) for downloading premium MMS applications and greetings, and 5 pesos for sending an MMS to another subscriber. These extra charges generate additional revenue to share with content providers and application developers.

Application Developer's Perspective

It is not surprising that application developers feel squeezed by the sheer number of players and pricing models currently available for launching MMS applications. The shortest and most stable path to the end user would be to make a distribution and revenue sharing agreement directly with a carrier. But it is very difficult for individual developers to get the attention of large carriers, and the smaller carriers will not have enough MMS customers to generate significant revenue for even the most popular applications in the short-term. Therefore, to reach the largest possible market, developers need a channel that will present their applications to more than one carrier at a time. Unless the application developer is also a creative artist, it is likely that prior partnerships with content owners will be required in order to populate the application with appropriate content.

As described previously, the application developer has some important bargaining chips to bring to the table. MMS tool makers and software providers are already proliferating, and there are almost innumerable online and traditional multimedia content owners looking for new revenue opportunities. But truly innovative MMS application ideas are still in short supply. Carriers understand that predictable applications like MMS greeting cards and photo albums are good enough for launching an MMS service, but these will not keep customers coming back for more, nor will they build traffic to the desired level. The relative scarcity of skilled MMS developers and applications, combined with the competition among value chain participants for innovative offerings, means that a number of paths are available to developers for introducing their applications to the market. These multiple options also provide an opportunity to negotiate more favorable revenue sharing and distribution terms.

Getting Applications to Market

Developers with compelling content and creative MMS applications do not have to look very far to find potential distribution partners. There are dozens of new companies claiming the role of MMS content aggregators, and many of these contenders have forged agreements with multiple operators. MMS infrastructure providers, including well established players such as MMSC vendors Logica/CMG, Comverse, and Ericsson, offer application partnership and distribution programs. Nokia, which manufactures both MMS handsets and MMSCs, has announced its own application partnership and distribution program. Openwave, the reincarnation of Phone.com of WAP fame, is positioning itself as an MMS storage solution provider, and is also sponsoring a developer's program. The BREW initiative, launched by Qualcomm to distribute and monetize handset-based applications for wireless operators, has now embraced MMS applications as part of its mission.

With so many partners available, the challenge is to find the best match for a particular developer. It is helpful to start the process by establishing the evaluation criteria that are most important for the short- and long-term success of the application. Some of the most important criteria for selecting a distribution partner include:

- Established carrier connections
  - How many carriers does this partner currently represent? What types of agreements are in place with these carriers? Is this number expected to grow significantly over the next six months?
- Commitment to individual developers
  - What level of support and promotion do the partners offer to you?
  - What is the quality of their marketing materials, Web site, and support documentation? Are development partners and applications featured?

  - Is there evidence that this partner has already worked well with developers of your size and scope?
- Costs and interface requirements
  - Is there an up-front fee for joining a developer program or testing and distributing your applications?
  - Are the benefits of the program commensurate with the fees?
  - How much effort will be required to interface with the partner's infrastructure? Is this a one-time effort, or will each application require new work?
- Business and revenue sharing terms
  - Will you receive an up-front payment for your applications, or agree to share revenue when it is received from carriers?
  - What percentage of revenue will you receive and what will be deducted before your share arrives?
  - Will you share in any extraordinary payments received by the partner for your application (sponsorship fees, special promotions, etc.)?
  - Are there any limits on making parallel agreements with other partners for the same application or content?

Comparing the answers to these questions will help the application developer understand the full value of the partnership, as well as create realistic expectations about the revenue that any distribution agreement may generate. While the larger and better-established partners are likely to have more fixed distribution terms, there is often room for negotiation, especially if a particular application has a demographically desirable target market and a strong business case behind it.

Matching Pricing and End User Value

Making the best possible distribution deal is a necessary step for selling applications, but the final decision to pay for a particular MMS message or service is ultimately in the hands of the individual consumer. Therefore, it is also important for developers to keep the end user in mind. Does the application appeal to a specific market segment (adolescent girls, young professional men, sports fans, etc.)? If so, how many mobile subscribers populate that segment today, and how likely are they to have purchased MMS capable phones? What will motivate the target customer to use your application and how often will he use it? Does the application have a unique value that will make it a candidate for premium subscription pricing? Will the original buyer want to share it with his friends? If it is a content-dependent application, do you have a flexible agreement with the content owner to accommodate demand for faster content turnover or customized versions for sponsors? All of these questions will impact the price that the end user is willing to pay, and how much of that price will come back to you in the form of revenue sharing.

Suppose that you have developed three different MMS applications that are aimed at different markets, and that each application has features that lend themselves to different pricing structures and partnership agreements. We will use this scenario to illustrate different revenue models and net revenue outcomes for the developer.

The first application lets customers view an MMS weather map for any city, annotated with current weather conditions. The customer sends an SMS with the name of the desired city, gets back a message confirming the location, and then downloads the weather map in MMS format. The cost to the customer is the same as that for any MMS, and the carrier shares this revenue with the application developer, who in turn shares 40 percent of it with the content provider, in this case a Web site that generates the weather maps. The market for this application is a subset of the devotees of online weather services.

The second application is aimed at customers who have purchased an MMS phone with a camera attachment and who are looking for a quick and easy way to customize their photos before sending them to friends. The customers will pay a one-time charge to download this photo editing application onto their phone, and the carrier or other service provider offering the application will share the revenue for each download with the developer. To get their applications to market, the developer has signed an agreement with an application aggregator, and this partner has a general agreement with the carrier. The aggregator agreement includes tracking application downloads and collecting monthly revenues from each carrier who signs on for the applications. In exchange for these services, the aggregator takes 20 percent of the revenues due from the carrier. That leaves 80 percent for the developer. Since there is no content partner involved, the developer does not have to share this revenue further downstream.

The third application provides subscribers with a fast new way to download and play exclusive backstage, preconcert images and interview sound bites, along with onstage performance shots and music excerpts for popular music stars. It is aimed at fans worldwide, particularly the desirable teen and young adult markets. Because this is a hypothetical example, also suppose that the developer has managed to sign content distribution agreements with several rising stars and has the in-house capability to feed appropriately formatted content directly to the MMS infrastructure of participating carriers. This application has enormous potential to generate demand for MMS services, so the developer is able to make a very favorable revenue sharing deal with the carriers, who agree to take only 30 percent of the revenue for the application itself and the follow-on MMS traffic it generates. On the other hand, the developer has agreed to turn over the majority of his net revenue to the content generators downstream—in this case, the musicians and their agents will get 75 percent.

As Table 9.2 illustrates, these different application properties (and the deals that the developer makes to bring them to market) will have a significant impact on the revenues that come back to the developer each month, even when the number of end users and the amount of MMS traffic is similar across applications.

TABLE 9.2

Application Pricing Alternatives and Impact for Developer Revenues

	1) Revenue Sharing Based on Fixed MMS Cost per Message	**2) Revenue Sharing Based on Premium App with One-Time Download Fee**	**3) Revenue Sharing for Premium App Download Fee and per MMS Message Fees**
Price to Customer	.50	2.50	3.50
Number of Downloads /Month	25,000 MMS	25,000 app. users	25,000 app. users 125,000 MMS
Carrier % of Revenue	50%	40%	30%

(continued on next page)

TABLE 9.2

Application Pricing Alternatives and Impact for Developer Revenues (continued)

	1) Revenue Sharing Based on Fixed MMS Cost per Message	2) Revenue Sharing Based on Premium App with One-Time Download Fee	3) Revenue Sharing for Premium App. Download Fee and per MMS Message Fees
Balance per Use for All Downstream Partners	.25	1.50	2.45 per app. download; .35 per MMS
Downstream Revenue Division:			
App. Developer %	60%	80%	25%A
Distributor %	N/A	20%	N/a
Content Provider %	40%	N/A	75%
Net $ to Developer per App. Download and/or Message	.15	1.20	.6125 (per app.) .0875 (per MMS)
Net Monthly Total $ to Developer	$2,250	$30,000	$15,312.50 (apps.) $19,937.50 (MMS)

Real-World MMS Distribution Paths

Until some winning business models for MMS distribution emerge from the current MMS value chain, developers will find that the real-world negotiations for MMS application distribution are even more complex than the alternatives described in Table 9.2. But developers are not entirely on their own in finding a path to the market. The following examples describe how other value chain players, including carriers, infrastructure providers, and application aggregators are working to collaborate with developers worldwide.

Bringing Developers on Board in Singapore

Advance planning for Singtel's 2002 launch of MMS services in Singapore went beyond the predictable advertising and public marketing campaigns. Singtel managers wanted to ensure that their MMS launch featured plenty of innovative applications and services so that customers who signed up and bought MMS phones would have a strong incentive to send MMS messages regularly. That meant getting application developers involved early, to make sure that the best application ideas were tested and fully functional before the launch date.

But there were some barriers to accomplishing this goal. While Singapore was already home to a community of mobile application developers familiar with SMS and WAP technology, MMS required a new set of skills and development experience. And because there were few MMS-capable handsets in the prelaunch market, and no test network for developers to access, the support structure needed for developers interested in learning MMS techniques was lacking. Singtel needed some way of jump-starting the MMS development effort to create a core of skilled and committed application developers who had a strong connection to the emerging MMS marketplace.

Given the complexity of the MMS value chain, it is no surprise that Singtel management decided to work with other value chain participants in building up the MMS development community in Singapore. As a leading MMS infrastructure provider and developer of MMS capable handsets, Ericsson emerged as an ideal partner for this collaborative effort. Engineers at Ericsson's Singapore-based Research and Development Cyberlab were already familiar with the MMS development environment, and Ericsson worldwide could provide easy access to deep MMS expertise, as well as an infrastructure and handset test bed. This combination provided a solid set of resources for developer training in MMS. The two organizations agreed to collaborate on a one-year program to encourage MMS development through business and technical seminars, hands-on training and testing facilities, and the potential of having completed applications adopted for commercial distribution by Singtel.

More than 100 interested developers participated in the MMS training programs sponsored by Singtel and Ericsson during 2002, and dozens submitted applications as candidates for inclusion in Singtel's MMS launch. As a result, Singtel was able to showcase MMS services ranging from personalized camera phone-generated "photo albums" to an MMS-linked traffic camera that monitors busy intersections and sends up-to-date traffic images to MMS-enabled commuters.

Even the program participants who did not manage to sign MMS application deals with Singtel emerged from the program with a much clearer understanding of the challenges of MMS development, along with the skills to integrate MMS into their ongoing development plans. Anthony Tse, applications solutions manager for Ericsson in Singapore, notes that, "Many of the participating developers came away with a commitment to MMS and a good understanding of how to implement their application ideas. As the MMS market in Singapore continues to expand, the development community is able to keep pace with demand for innovative apps."

Adding MMS to the BREW

The intensive training provided by Ericsson and Singtel in Singapore is not, however, available to the majority of developers interested in testing their MMS applications and finding customers. One example of a more general program for application distribution and monetization is Qualcomm's BREW developer program. BREW has a well defined development interface and testing regimen. Once developers have an application emerge from the testing process as "TRUE BREW"™ compliant, that application can be downloaded onto BREW-compatible handsets. But, unlike many development programs, BREW also provides support and a clear framework for the carrier to charge for application downloads and for the application developer to share revenues. For each paid application download, the developers receive the majority of the revenue generated, while the carrier and Qualcomm take a smaller portion for providing the application download service.

The BREW application program has boosted the ability of code division multiple access (CDMA) carriers worldwide to get applications to their customers. BREW is air interface independent (i.e., it can work on GSM, GPRS, IS-95A, CDMA2000, WCDMA, etc. networks), and Qualcomm is prepared to BREW-enable all wireless technologies. BREW sees itself as providing much-needed application testing to augment the carriers' established interoperability and handset testing programs. This helps give the carriers confidence that all BREW-tested applications will work as expected across the network. While CDMA2000 carriers and infrastructure providers have not been as aggressive as GSM and GPRS networks in adopting MMS standards, the BREW model provides a clear path for application developers to get their applications in front of carriers. Once they sign up for the BREW program, developers can request that the program test their applications on a particular carrier's MMSC connection, helping to establish that the application will work as expected on that carrier's network and BREW handsets. BREW also enables developers to create an MMS extension, which is a piece of client software for the handset that provides a standard MMS API for different types of MMS content from one or more developers.

Paul Burlingame, senior product manager of Qualcomm Internet Service, notes that the MMS BREW extension concept will give developers and carriers a great advantage in speed to the market. Instead of having to track and adjust to all the different specifications for new MMS handsets, they can write their applications based on a stable BREW MMS extension that will be common to different handsets. This approach will also allow the carriers to update MMS capabilities over the air as needed, avoiding the expensive process of having subscribers bring their handsets back to a dealer for upgrading.

Burlingame also notes that BREW carriers in China and the United States are particularly interested in MMS applications. This is good news for developers looking for mass-market distribution.

Summary

Business models for the distribution and sale of MMS applications and services are complicated and still shifting. It is likely that one or more of the roles in the current MMS value chain will become redundant or be consolidated into a larger set of services. Therefore, it is a safe prediction that there will be a shakeout among the growing number of MMS toolmakers, infrastructure providers, portal services, and so on.

There is also a threat that the scramble of too many players for pieces of the MMS revenue pie, combined with too little focus on the value of applications and content from the customer perspective, will lead to overly complex and costly services. This outcome will slow down MMS adoption, resulting in less traffic than projected and limited opportunities for all players.

What is the best course for application developers? Understand the business options, but focus on delivering value to the end user. Define your application ideas and implementations in terms of this value, and focus on partnerships that bring you closer to the customer.

CHAPTER 10

Next Steps in Multimedia Messaging

The potential of multimedia messaging service (MMS) and its impact on mobile application revenues, like the prospects for every new technology, are dependent on the fulfillment of a number of technical and business assumptions. Projections for the worldwide growth of MMS are, in the parlance of the typical new stock prospectus, "based on forward-looking statements that involve market risks and uncertainties including, but not limited to, the ability to develop and commercialize the products under discussion." As noted in earlier chapters, application developers have little control over how the market ultimately judges the worth of MMS, even though issues such as MMS interoperability, roaming agreements, and business models will determine the return that those developers may get from any investment in launching MMS applications.

Developers cannot build their applications based on speculation, however. They need to wrestle with the here and now limitations of today's MMS formats, handsets, application management infrastructure, and available network connections. The balance of this book, therefore, has been firmly grounded in the reality of MMS development, as of 2003. This final chapter shifts focus to consider the major issues that will either propel MMS to reach its full potential or stand in the way of the rosy projections about future killer applications and customer adoption rates.

Lacking access to a crystal ball, we will not try to predict how MMS will look five years from now. We expect it will take at least that much time before the full trajectory of MMS becomes visible. Meanwhile, it is worth taking a closer look at the issues that application developers and MMS infrastructure providers will be watching closely.

Issues to Watch

Content Protection and Digital Rights Management

If, as expected, a significant percentage of future MMS transactions include brand-name content, then content owners will want evidence that their property will be protected from unauthorized copying once it gets into mobile distribution. High value multimedia content owners are not interested in any new channel that looks too much like the open access model of the Internet. So, along with the arrival of content at the mobile applications table comes a new concern about the transmission of content bits across networks. MMS content agreements are encrusted with worries about where these content bits will go, how long they will stay alive, who handles them, and what middleware and mobile network gateways might do to them along the way to the end user's handset.

The future availability of MMS content depends on addressing these concerns, while simultaneously expanding the options for digital rights management (DRM). Customers who pay per message to view MMS content will not be satisfied with one look only. Imposing such tight content protection requirements that paying customers don't get to use their MMS messages as they wish is a sure way to cut off market demand at the outset. Agreements among content owners and distributors need to support controlled storing and forwarding, and even copying of MMS content. Some of these capabilities will be built into the basic service, and others will probably require some small additional payment.

The thinking about DRM for MMS among wireless carriers and distributors tends to work within a very narrow and focused definition of this problem. To most mobile industry players, the only important measure of a DRM solution is how successful it is at preventing unauthorized copying. In some cases, this concern gets so overpowering that all copying is suspect, and only the most demonstrably benign forms of local use are to be allowed. This is a dysfunctional approach to the problem of DRM for two reasons.

First, copying material with attached digital rights is good for everyone concerned if it is properly reimbursed. Blockbuster sells copies. Movie houses play copies. Radio stations broadcast copies. Without copies, the only existing content would be originals hanging in museums. Copies are how returns are multiplied.

MMS value chain players should not aim to prevent all copying of MMS content. Instead, the goal should be to combine control over copying with an easy and immediate way for the end user to pay for the level of copying that is desired. This could include carriers or

MMS infrastructure providers charging a transaction fee and managing the reimbursement for copying. Mobile subscribers have already exhibited a willingness to pay for content and applications downloaded onto their handsets that far exceeds market acceptance of fee-for-content on the Web. If MMS players make full use of that willingness by offering subscribers copying options that are easy and low cost, they will multiply revenues instead of creating a black market industry of tools and techniques to circumvent restrictive MMS copy protections.

An equally important content and copy protection issue is ensuring the quality and integrity of the original content as it moves over multiple wireless networks. In the process of transcoding and adapting an MMS message content for a particular handset, it is not acceptable to change the color of a graphic just to make more efficient use of that handset's color palette. In transcoding and content adapting an MMS message for a particular handset, it is not acceptable to crop the image or resize it without maintaining the aspect ratio. Nor is it acceptable to cut off the last three seconds of the audio track. Nor is it acceptable to drop two channels from the musical instrument digital interface (MIDI) track. Nor is it acceptable to transform the Wave (WAV) file into an adaptive multirate (AMR) file.

These are much more than mere technical details to content owners. They care deeply about the quality of the content because their reputation and livelihood may depend more on their original product retaining its precise graphic and sound qualities than on traditional forms of copy protection. Until MMS content can provide this type of quality assurance, as well as subscriber-friendly copy protection and copy payment options, it will be problematic to bring high-quality content into the value chain.

Interoperability of Hardware and MMS Standards Conformance

The telecommunications industry is fanatical about standards conformance when it comes to homogenizing and commoditizing the wireless network products it buys. The network has to work end-to-end, and costs must be minimized. Wireless carriers, therefore, have worked hard to avoid interoperability problems for themselves by limiting the adoption of proprietary features in their network architecture and pressuring vendors to conform to standards.

Developers, like mobile subscribers, don't have this type of clout when it comes to mandating interoperability and standardization of the human and development interfaces presented to them. As anyone who has tried to configure a wireless application protocol (WAP) phone or set up a general packet radio service (GPRS) connection will testify, the problem of making all the user-facing pieces of the network work successfully is typically left as an exercise for the subscriber. In fact, it seems that the mobile service and device providers make it as difficult as possible to do anything outside of the plain vanilla setup or the "walled garden" of services and content provided directly by the carrier.

Telecommunications standards in general, and European Telecommunications Standards Institute (ETSI) standards in particular, have been spectacularly successful in getting the core network to work. Unfortunately, early indications are that the same carriers and handset manufacturers who are launching MMS are failing to adopt a rigorous insistence on standard, interoperable interfaces among devices and networks. A telling example is the MMS Conformance Document, a joint document published by the leaders in MMS infrastructure including CMG Wireless Data Solutions, Comverse, Ericsson, Logica, Motorola, Nokia, Siemens, and Sony Ericsson. Early in 2002, these companies banded together to create an Interoperability Group for MMS. The stated objective of this group is ensuring the smooth introduction of MMS into the market, including "seamless end-to-end operability between MMS handsets and servers from different vendors."[1]

In accordance with this mission, the MMS Conformance Document authored by the group seeks to provide a clear definition of the minimum set of requirements that all MMS handsets will support. This document could and should be an essential resource for MMS developers who need to test their applications across different

---

[1] Ericsson Press Releases, "Leading telecommunications companies cooperate on MMS interoperability," February 19, 2002 (http://www.ericsson.com/press/20020219-075611.html).

phones. Unfortunately, strict compliance with the Conformance Document during the development process does not result in applications that are interoperable among the handsets manufactured by the members of the Interoperability Group. Apparently, each manufacturer has already added proprietary extensions and capabilities to its handsets that make the Conformance Document far less valuable.

For MMS to be a worldwide success, interoperability and standards conformance have to be more than popular phrases for press releases. Neither the application developer community nor the wireless subscriber will embrace a technology that comes packaged with frustrations and inconsistent performance. The same insistence that wireless carriers have brought to bear on network standards is needed in the next stage of MMS evolution.

Roaming Between Operators and Between Network Technologies

The difficulty and expense of sending MMS messages between different carrier networks is another challenge to MMS adoption. MMS roaming, or more properly the pursuit of MMS roaming charges, is much higher on the industry's agenda than is compatibility among handsets. What we may find, however, is that like voice services, the two cannot be decoupled. MMS roaming may, at the end of the day, force certain homogeneity or at least commonality across the MMS platform. There are two reasons for this.

First is the issue of transcoding. Transcoding is one point where 3GPP and 3GPP2 differ. The 3GPP folks tend to sweep every interoperability problem that comes up under the transcoding rug. The 3GPP2 folks tend to make minimizing the need for transcoding an explicit requirement of their designs. If 3GPP and 3GPP2 want to exchange MMS messages and 3GPP2 does not transcode, then the handsets will have to start to look the same.

The second issue is the matter of digital rights management. From the point of view of the network operator, voice services are by and large fire-and-forget. The operator picks up a call, drops it at a connecting point, and does not think more about it until the billing record arrives. Digital rights must be tended to end-to-end. If the sender sets the "Don't Copy" bit and hands the MMS to the local operator, he expects that the bit will still be set, and even more, will be obeyed when the MMS gets to the far end, wherever that may be. Dealing with this end-to-end property of MMS, like the transcoding property, may tend to control the amount of gratuitous difference that grows in the system.

Value-Based Pricing for MMS Applications

From the service provider perspective, transcoding fees, roaming fees, possibly DRM and copy protection fees, data storage fees, and of course MMS message fees, may all look like little pots of gold hidden in every nook and cranny of the MMS architecture. But if every participant in the MMS value chain insists on collecting a coin every time an MMS message moves to a different segment of the end-to-end architecture, the typical mobile subscriber will be less and less eager to send and receive MMS messages. Delivering value to the subscriber population must be the first order of business for everyone with a stake in MMS adoption.

At this early stage, it is hard to know what will become the "killer" applications for MMS, but there is no mystery about what mobile subscribers generally value. This list includes:

- Ease and dependability of use.
- Predictable pricing and billing.
- Peer to peer communication capabilities across networks.
- Service and customer support.
- Variety and choice in services and applications.

In addition to providing true MMS interoperability and consistent roaming services, carriers need to work with the other MMS value chain participants to ensure predictable pricing and billing for MMS services. A combination of fixed fee per MMS message, along with a small additional charge for downloading premium applications or brand-name content is a good start, as long as subscribers don't find unexpected charges for other MMS fees on their monthly bill.

Premium MMS content and applications of all kinds also have to be freely available to subscribers across carriers. The "walled garden" approach, in which carriers and handset makers limit the content and services available to their subscribers, is a misguided approach to the market. In addition to preventing subscribers from exploring and embracing new services, it distorts the appropriate market price for premium content. When carriers lock subscribers' phone menus to display only the content and applications provided by telecommunications insiders and paying advertisers, they trade short-term revenue for long-term demand and market growth. Particularly in the early stages of MMS development, it is essential to give customers a choice about what goes onto their phones and what is worth paying for, if for no other reason than to discover what these services are. The only way to accomplish this is to keep the channel open for developers and content owners to display their wares on the phone.

The Role of the MMS VASP

An animal of unknown dietary preferences in the MMS food chain is the value-added service provider (VASP). In particular, it is too early to tell if the typical MMS VASP will be inward looking and dancing to the tune of the network operator, or outward looking and catering to the needs and demands of application developer and content owner customers. The first generation VASPs we see today look much more like the first type of animal than the second. Like short messaging service (SMS) brokers and mobile application certification services, the VASPs are attempting to serve the needs of network operators and become a conduit between the carriers and MMS application developers.

If the VASP is just a taxman, then this will be just another barrier for the MMS application developer to overcome, and another opportunity for the MMS application developer to wonder if there is a better way to try to make a living.

If, however, the VASP faces the MMS application development community and concentrates the revenue-generating potential of that community, along with the community's network traffic, then the MMS marketplace will look more like the best of the ringtone download market—which so far has generated more revenue for content providers, developers, and carriers than all of the J2ME and BREW mobile applications combined.

Interaction with Technical Substitutes: Polyphonic Ringtones, EMS with SVG, IMS

The thriving SMS messaging and application market and the lucrative ringtone download market prove that it is indeed possible to generate billions of dollars from relatively simple value-added mobile applications. Both these markets are highly profitable for the network carriers and for a segment of the content and application providers. SMS offers the subscriber instant and uncomplicated gratification in a messaging paradigm. Polyphonic ringtones and enhanced messaging service (EMS) with scalable vector graphics (SVG) add better audio and dynamic graphics, respectively.

One school of thinking about MMS—the messaging school—takes the success of SMS and mobile messaging as a conceptual model. From this perspective, polyphonic ringtones and EMS with SVG are the dying gasps of their respective markets. MMS will leap over them technically and catch the next exponential messaging curve with hot videos and MP3 audio. MMS is Internet chat merged with MTV, mobile, and voice.

The Internet protocol multimedia system (IMS) integrates the mobile handset with the Internet; it converges all of the operator's fixed and wireless services into one grand, unifying calling plan. It is two-way, always-on WiFi with device portability and great coverage.

The other school of thinking about MMS—the multimedia school—takes the Internet and surfing as its model. According to this perspective, WAP failed not because Web surfing is a mismatch for mobile devices, but because early WAP was black and white, lacked graphics, and was too slow. Now we have 3G data, multimedia handsets, and Internet connectivity. MMS is a Web site with time added.

Both the messaging thinkers and the multimedia thinkers miss the point. (The multimedia thinkers have a distinguished history of missing the point, but that's, well, beside the point.) There is a communication component of MMS, but it is not person-to-person like SMS. And while the sensory channels of today's mobile subscriber are certainly underutilized, but there is nothing in the architecture of MMS that turns the phone into a wireless Windows Media player.

We have tried to argue that MMS is something new. This is its strength and its weakness. It is not SMS nor is it orchestrated scraps of World Wide Web on an air interface. Polyphonic ringtones and IMS tell MMS what it is not and should not try to be. They don't tell MMS what it is.

Ringtones, EMS, and IMS are not technical substitutes for MMS. And they are only threats if MMS tries to be like them. The overarching challenge for MMS is to find its voice. Its voice, if it is to be heard, is in the street with the early adopter subscribers and in the home-grown servers of the independent application developer. It will not be found in the marketing department of the network operators or the infrastructure manufacturers.

Expanding to New Markets

Sponsored MMS and MMS Advertising

More than any other mobile application format, the structure of the MMS message and its display on the phone screen lends itself to featuring the names and logos of sponsors without interfering with the quality of the message itself. This opens up a number of sponsorship and advertising opportunities. Sponsor-subsidized MMS services, like the sports highlights described in Chapter 1, are a natural way to reduce the cost of multimedia on the phone and to create a consumer demand for premium content. In addition to motivating subscribers to try MMS, the promotional campaigns that accompany sponsored programs will provide another rationale for consumers to upgrade to multimedia handsets with integrated cameras. And once a critical mass of consumers own the handsets, network effects will kick in and make it more attractive to use those handsets to exchange MMS messages with others.

So far, there have been only a few notable sponsorship programs featuring MMS. Coca-Cola, for example, sponsored the launch of MMS messaging by T-Mobile in Austria, paying for subscribers to send each other messages with Coke themes and graphics. Christmas celebrations in Finland in 2002 included free MMS mobile greetings, music, and seasonal graphics sponsored by one of the wireless carriers, DNA Finland, in cooperation with Joulupukki TV, the producer of Finland's Santa Claus Internet television channel. These early examples of sponsored MMS services provide proof of concept and initial subscriber acceptance of advertising messages on the phone, but they are a far cry from the mass marketing campaigns needed to generate significant advertising revenue.

Consumer acceptance will be the key to growth of MMS sponsorship and advertising campaigns. The omnipresence of banner and pop-up advertising on the Web, combined with floods of e-mail marketing spam, give online advertising a bad name. Wireless subscribers are understandably wary of opening their mobile phones to a similar flood of unwanted messages. That means strict and convincing opt-in and termination policies for any MMS marketing, along with clear control by the consumer over how and when marketing messages display on the handset.

Enterprise MMS

Carriers are looking to the consumer market to adopt MMS and generate high-volume traffic from content and messaging. The enterprise demand for MMS applications has not had nearly as much attention. However, MMS is well suited to a number of enterprise applications, including communication among employees, corporate-to-consumer messaging, and business-to-business applications. At least one analyst group, Juniper Research, projects that $1.4 billion of MMS revenue will come from enterprise customers by 2004.

What types of MMS services and content will generate this revenue? Another Finnish company, GISnet, has announced an MMS application for fieldwork management called Job Dispatcher. This application is linked with online mapping and location resources to enable corporations to pinpoint the location of field workers via their mobile phones, determine the most efficient route for the worker to get to his next appointment, and then transmit a map and directions to the mobile phone.

Other possibilities for MMS-based fieldwork include using a camera-enabled phone for transmitting images of everything from accidents and damages back to insurance companies, to damaged parts of field equipment back to a maintenance depot, to diagnose mechanical problems and automatically order replacement parts. Once MMS phones become more widely available, companies could also provide operating instructions and troubleshooting information via MMS for products that are typically used out of the range of computers. This type of troubleshooting service might start with diagnosing malfunctioning snowmobiles in Finland and eventually spread to outdoor appliances and equipment of all types.

Enterprise adoption of MMS, like the adoption of other mobile enterprise applications, will depend on a demonstrated return on investment, as well as tools to integrate MMS messaging with existing IT architecture. One key driver, therefore, will be the integration of MMS into the products of enterprise information technology solution providers. Application developers who want to target the enterprise customer with MMS services will need to seek partnerships among this group. It is unlikely that wireless carriers or mass market MMS infrastructure providers will take a lead in providing enterprise-focused MMS services.

Conclusion

The success of MMS—if it is to be—will flow from a combination of the issues described in this book, and no doubt from some additional technical and business factors that no one can predict from this early development perspective. Fortunately, MMS today is already working well enough to yield countless opportunities for committed developers to create real value for customers and distribution partners. We hope that this book contributes to the growth of innovative applications and that it brings MMS one step closer to reaching its full potential.

APPENDIX A

MMS Standards and Specifications

The standards and specifications defining the multimedia messaging service (MMS) can be roughly divided into two categories: network and media.

Standards and specifications in the first category focus on the network protocols for carrying the multimedia messages. These documents take into account the nature of the messages and their expected usage scenarios, but they do not delve into the encoding of the multimedia message itself. The network standards and specifications are maintained primarily by the following organizations.

- 3GPP at www.3gpp.org
- 3GPP2 at www.3ggp2.org
- Open Mobile Alliance at www.openmobilealliance.org
- Internet Engineering Task Force at www.ietf.org

Standards and specifications in the second category concern the encoding of the multimedia message itself, and not how it is moved from one point to another. These standards and specifications take into account the fact that the message is moving through a low-bandwidth wireless communication channel, and that it is being rendered on a small, handheld, battery-powered device with only modest playback capabilities. By and large, each media standard and specification is maintained by an industry group that is focused on that particular media format.

Network Standards and Specifications

3GPP TS 22.140: "Multimedia Messaging Service: Service Aspects—Stage 1"

A relatively short document compared to the others, but a critically important document. TS 22.140 is the requirements document for MMS. It says what MMS should be able to do and what MMS does not do. It defines the service. By the rules of 3GPP, a new feature or capability can only be added to the system if it is added to the Stage 1 document first. If you want to know what MMS is and will be, track this document.

3GPP2 S.R0064-0: "Multimedia Messaging Service (MMS): Stage 1 Requirements"

This is the 3GPP2 version of the preceding document; that is, MMS requirements from the 3GPP2 point of view. Since almost everyone wants to achieve interoperability between 3GPP MMS and 3GPP2 MMS, it draws heavily on TS 22.140, but strangely does not reference it.

3GPP TS 23.140: "Multimedia Messaging Service (MMS): Functional description—Stage 2"

This is the key document for understanding MMS from a technical perspective. It is the high-level, overall technical architecture of MMS with some details of the interfaces. It defines the technical modularization, interfaces, and vocabulary of the service. It is a good idea to find out where you are in its diagrams.

3GPP TS 24.011: "Point-to-Point (PP) Short Message Service (SMS) Support on Mobile Radio Interface"

MMS often uses SMS to send the MMS notification messages to the end user.

3GPP TS 23.060: “General Packet Radio Service (GPRS): Service description—Stage 2”

This is more than you will ever want to know about GPRS. But if you do ever wonder what is down there getting your MMS to the phone, this is the place to start.

WAP-205-MMSArchOverview-20010425-a: “WAP MMS Architecture Overview”

This document describes how the MMS architecture in 3GPP TS 23.140 fits into the WAP architecture. The division of work and responsibility for the evolution of the MMS architecture between 3GPP and OMA is fuzzy and fluid.

WAP-230-WSP: “Wireless Application Protocol: Wireless Session Protocol Specification”

This document describes how generic data types such as strings and integers are WSP-encoded. It also includes binary encodings for multipart multipurpose Internet mail extension (MIME) content.

WAP-209-MMSEncapsulation: “WAP MMS Encapsulation Protocol”

This document is a companion to WAP-230 that defines WSP-style binary header encodings for MMS messages.

Some headers are defined in both the MMS encapsulation protocol and the WSP specification. The same text header can have different binary representations based on whether the message part you are encoding is governed by the MMS encapsulation protocol or WSP encoding. The MMS encapsulation protocol is used for the header of the binary “application/vnd.wap.mms-message” content. WSP encoding is used for hypertext transfer protocol (HTTP) header encoding performed by a WAP gateway, and for the binary encoding of multipart messages, such as the payload of an MMS message.

Media Standards and Specifications: MMS-Specific Usage

3GPP TS 26.140: "Multimedia Messaging Service (MMS): Media Formats and Codecs"

This document lists the media formats that shall or may be supported by an MMS. It is an excellent place to track the arrival of new media formats. The reference section contains up-to-date URL pointers to all the detailed media format specifications supported in MMS.

3GPP TS 26.234: "Transparent End-To-End Packet Switched Streaming Service (PSS): Protocols and Codecs"

There is generally an effort to use the same media formats across all mobile services in 3GPP. The definition of those formats has not yet been factored out of the documents describing individual services, but at least the individual service documents are pointing to each other. This document contains the definition of 3GPP synchronized multimedia integration language (SMIL) that is used in MMS.

Media Standards and Specifications: Individual Media Formats

"The Unicode Standard," The Unicode Consortium

ANSI X3.4, 1986: "Information Systems: Coded Character Set 7 Bit—American National Standard Code for Information Interchange"

IETF RFC 2046: "Multipurpose Internet Mail Extensions (MIME) Part Two: Media Types"

IETF RFC 2083: "PNG (Portable Networks Graphics) Specification, Version 1.0 ," T. Boutell, et al., March 1997

IETF RFC 2279: "UTF-8: A Transformation format of ISO 10646"

IETF RFC 3267: "RTP Payload Format and File Storage Format for the Adaptive Multi-Rate (AMR) Adaptive Multi-Rate Wideband (AMR-WB) Audio Codecs "

ITU-T Recommendation T.81: "Information Technology: Digital Compression and Coding of Continuous-Tone Still Images: Requirements and Guidelines"

ITU-T Recommendation H.263: "Video Coding for Low Bit Rate Communication"

ITU-T Recommendation H.263 (1998): "Video Coding for Low Bit Rate Communication—Annex X, Profiles and Levels Definition"

ISO/IEC 8859-1:1998: "Information Technology: 8-bit Single-Byte Coded Graphic Character Sets—Part 1: Latin Alphabet No. 1"

ISO/IEC 14496-2 (1999): "Information Technology: Coding of Audiovisual Objects—Part 2: Visual"

ISO/IEC 14496-3:2001: "Information Technology: Coding of Audiovisual Objects—Part 3: Audio"

"GIF Graphics Interchange Format: A Standard Defining a Mechanism for the Storage and Transmission of Raster-Based Graphics Information," Compuserve Incorporated

"Graphics Interchange Format (Version 89a)," Compuserve Incorporated

W3C Working Draft: "Scalable Vector Graphics (SVG)"

W3C Working Draft: "Mobile SVG Profiles: SVG Tiny and SVG Basic"

W3C Recommendation: "Synchronized Multimedia Integration Language (SMIL 2.0)"

3GPP TS 26.090: "AMR Speech Codec Transcoding Functions"

3GPP TS 26.071: "Mandatory Speech Codec Speech Processing Functions: AMR Speech Codec—General Description"

3GPP TS 26.171: "AMR Speech Codec: General Description"

RP-001: "Standard MIDI Files," MIDI Manufacturers Association

RP-34: "Scalable Polyphony MIDI Specification," MIDI Manufacturers Association

RP-35: "Scalable Polyphony MIDI Device 5-to-24 Note Profile for 3GPP," MIDI Manufacturers Association

"JPEG File Interchange Format," Version 1.02, September 1, 1992

APPENDIX B

A Complete MMS Message on the MM1 Interface

```
X-Mms-Message-Type: MM1_submit.REQ
X-Mms-Transaction-Id: 1234-ABCD-5678
X-Mms-3GPP-MMS-Version: 6.0.0
X-Mms-Recipient-Address: 16177929194/TYPE=PLMN
X-Mms-Subject: My First MMS
X-Mms-Sender-Address: sguthery@mobile-mind.com/TYPE=RFC822
X-Mms-Delivery-Report: Yes

Mime-Version: 1.0
content-type: multipart/related; boundary="_----------
      =_103140636136200";application/vnd.mms.multipart.related;
      start=<1000>;
                                        type=application/smil
Date: Sat, 7 Sep 2002 13:46:01 UT

This is a multi-part message in MIME format.

--_----------=_103140636136200
content-disposition: attachment; filename="hello-world.smil"
content-transfer-encoding: 7bit
content-type: application/smil; name="hello-world.smil"
context-id: <1000>

<smil>
```

```
    <head>
        <layout><root-layout/>
            <region id="region1_1" top="0" left="0"
              height="100%" width="100%"/>
            <region id="region1_2" top="0" left="0"
              height="50%" width="100%"/>
            <region id="region2_2" top="50%" left="0"
              height="50%" width="100%"/>
        </layout>
    </head>
    <body>
        <seq>

            <par dur="3000ms">
                <img src="globe.gif" region="region1_1"/>
                <audio src="wooh.amr"/>
            </par>

            <par dur="3000ms">
                <img src="world.gif" region="region1_2"/>
                <text src="Text0000.txt" region="region2_2">
                    <param name="foreground-color"
                     value="#000000"/>
                    <param name="textsize" value="large"/>
                </text>
                <audio src="applause.amr" end="2550ms"/>
            </par>

        </seq>
    </body>
</smil>

--_----------=_103140636136200
content-disposition: attachment; filename="globe.gif"
content-transfer-encoding: base64
content-type: image/gif; name="globe.gif"
```

R0lGODlhrADNAIAAAAAAAP///yH5BAEAAAEALAAAAACsAM0AAAL+jI+py+0P

o5y02ouz3rz7D4biSJbmiabqyrbuC8fyTNf2jecXwOv+fwAEgcQcT1hM2pDK
Zkx4dEpXRyhzig31AlFu9vu5isFkTfWMTpfXi+7jyt2yyfIhpTrH1jfIfd7X
BReURohn4Pe3NIZQiBanFpeIg/gIGcnY+ChJQ1noteUIVem4+dIDh3d2OCbq
9TlqWEolCOt6uVrHJBYVK3viB2mF2Td4aDyKS+vr4TaY22zIK0q8ity83CGo
2lvpfCyn2ydMis13fepmBX6ZrjuE7lm+E5icflvtjnrK6CquriyvAbdGrbol
s3Xs3j9Q/wJGUJUgFa+DzogtUmgLniX+hwogDvMIMRWuiJj4jZNWkmNEj/VW
MtwnkaSdYtNksuQY01vLit5aAcyFkB0sSstI+WOJtB20d+/UfVr4FCe5Wjp3
7nTHNGHBb6+4+dqo6WmgaN3CzaSG1l7Fm6U06gMpcuRamf24FlsrkWgeqB8z
KkU4VWvVjB8H6l3Trq9YnvX8dbSzh9Xdxl7nECIZ1yBFa1g3z3xVrV+8P6HY
bYPrUyS6hPweTx5aM6xl1ZgbM6ZIzRgtgJ8XRr5W5uXv06h30Q7Kurffb5nY
RBteGO/Q1ltb0wX92ZpsOuCgFz+aOq6y3denId3+RanxWNvwwa6LPHlyn7pp
s50S6hlMggf++Xa+nh09bfCX3lj7UAaeb2WlNB4DqEQ1kVfANSENe7UkVaF6
oDjIoU2wZXhfEmp41x57JgrYSYdZrUZQiEUQ95aGR0k32Xjk1XZRaIo58dxw
wpnX3TmACaRidoKtE5iIjkFzGW6GHWjka1H20iJYRMCjmW1XnQgcXykBKN+H
BVX2g4FMhhMkkj5Gstp0Ob5RG2UUJijOdFZhiBKIToOW1Ur1tagEjDHeVhyb
2rhE1mJc5dZkdC/q+RYyC+o2IFBENiXULTn6lqQOgBY2450jgmndY1ttyBqn
VXY6i56KFdrShGFGmSo+i3g5jECs/uLWqeRwaRxdvA2bq6H0lWj+JZWyjgCp
fsIVi6CLpM6q4xuZWPklr3m5KhuwdXZUWqkTNCiWNonKBS5vzGiEHpaaPTlt
uVCKi2g+qwoY72HmEMiYenM55p4DvSJyzoQ1ARzfT+re4UmjYT3nF1A2+kkt
XvdePOXEHHR5mbNpVpUZYMJ0xZZZUho5cJViLqyrrGNleSK0Flo7EpbsHtqG
qU2dmzPFFWBr7KjPvgsshw3miWq29E7KM6n6ulTeiDehNld/A8NZL3LVzZrf
RA7uxvJg0mXYbqh9wiyprfEReWPP9e3sIsFh3yO2sZrE2F3EaOa9In2Xgl1k
36eB+/UdhIPt1HY9pomwZ7kR9veAbpv+lOy3ko8LtK0O5/Wn3qKJtuzakJvU
4dU0t8zwl2Y+C13aUwnKMrEVp5s2BHqFfjnXNd8L8+oytpdpkRLX+zTIiFtg
qWsrD+r4mZWd9PSDf4aEucZZYr708mdS5bqErzdcqsOFpxv2rrZL/zW/3Xut
7H3frbpD7Q+RiT2hNk5tmo/7Qakg7+sWT765CWxk6Drf3i6GQACaoUkCFN8C
FYY8EQhwHpnLnsVwNw9r6UuB48oGn+JHvAo+cH7hKkHs1CdB/phvXws7VAI/
pL1RpQCBWiJBiDiItdl1UAsJ/A4KAjPBHEoQAz3cVkikFQYH8uEXLSvi/gi0
HxZ8iog/dCL+/DBmtxZYkX4Z2CL4OOOW5dEkiNl4mBfPuMULzetB4OsR3VTQ
pq6hUYXFehJU7ijDEL6MiyaMoxwv+MKHfW4cF9Jjw4LBwMS5gJBTnONJQAca
8byHkaYCkUEqBANCrgyQYdzM8xgHJDZFEjOPCyPEMNjHkR3RY22UUaU4VxLg
WUSVEGoO90whyruZsiHQYxxNQCc0LqqKU5vMYiZzuThQDfNXEAsakABlsx8F
w4wEzGSb7MbKZfoLGLHp4BOZBExEYseakbTk2JqVFESJbm2MimFdElPNYx4I
aetJmbtiZc5L6cyLkLmmDCK0O3QaqiziPJLudPVGOLHLmP+8phv+/wU7DZFv
gAIL5xXhuU4tDvQ95+yYRyUnO9rRUWWam0QudekxGDYklgaqaEW3Frhk4tAE
B0OQOq/lO0rdjW7EqidJQWkERaLUkP1TVgBBNdES5gyR/qxBTZ1Z0kOKx46x
Eqnfl0mrnsy0iiixk9WkJqp+Ns979izrjLb6w532B2oqXGXimEo9Ohp0pZgs
01O7ykmyyWUdmuJex6iDRVGidYZdoZHmMDk072nHNojtlZh+OdgZuktqP/WW

yA7oNY72zI+QJeMi1XpL/xlFkdL06WVpuNPIzuKkeGJgtz55yhKpNHzU7KaI
CusZlTqvqci0RIRqGtElwfRKuL2kEXP+CkabEs+ZWPytHgB6veP2DpaD+xVz
h3TXnOhBUdPtZVENxljm1Mp/gs0occ1TQIJitoZIDdUZrWbeR3VztwDb1kZD
U9doLdW66A3OUxW3IYZQLr8somc2GxUTz3rqYN/qJEphSdRnwtW3UdzLf+n7
EqgmdSnHuSBP4ukcBhewOfF4maY8atYPJzTEukQXikssYItNMp+pjS8Yoqni
0jZTj6F97L9szJ2h9tWVQs3ri5m5v4DAxDQe1q00fUyc9iq4QHuClVvp+lGL
0necDmHR+nLaCanakjNA3kSBqeKlTh0ZLFGesoWvW6OH+m6W8jukShD6ZMaO
NkEwzNqS70z+sx8pM73RMtgpVwzow4FHqzdsdJsRnejCsZnPDw1ggiEdadSJ
L8G/SyR3M01EOjHP0qNlKKgfiNldta/Kp2YWkRsn6Hu2OpVrelfAUDlrEFA4
fXXLtSmc6Oug/rfXwW4oSYstbBQi26nfc/OyU1gyZz87idGetiLwJ21rbwyK
2uZEkpDYbRtaMdxc/earME1uCgLyqNVKtznwK69IZdjdXWzwMqFmanpLAFXe
BWJXVWvtOhFwivhEt74hm794F+yDB3/1I92JVfQ0nKWqCaxNcR3uqy18bzzG
eLelm+MgRXjiRK3jYopM8Inb1z3XArCy6W06CQdtuspVeSLnXOXMYNX84ArS
rNGcJWObk7aILh80z5v9Qg4bXd8bWVxbJYPVFc7a33zcGekiHq9cU11oJKsm
x3AacLL+lcFLcfiYi/3tNhbWoU08u69Z1XKy24PCbtc6Esdd6LIuW4Shthdq
0a7EbSfN59OuuwcH//G/u/pxgL+o1PelQyWLGacepyDj5TH2zGqz6v+j1Vd4
F+CUawu/4L5xMHv4hOENdzZ4lyz8eE3y6qVRTrHXoCMDH/vbH7v2RtZ9tnkP
/OALf/jEL77xj4/85CsfBQUAADs=

```
--_----------=_103140636136200
content-disposition: attachment; filename="wooh.amr"
content-transfer-encoding: base64
content-type: audio/amr; name="wooh.amr"
```

IyFBTVIKNEJpCKiMBZtOw8ADAaLtSD+HIbQAALtNtYAANIwjbToQJNlHStED
J/BiS+8MI+UXHlS1N8FQNLTIJRSAIXFO+GOcBLQxWyx4XIEVuomL4DAQNIVs
xMkAAO/HQixAO8iIF1uXISEYXr1lYeewNHEhJp2DJJ471b6eNVE3c9BNMLHe
xnZwOdUgNGlJMMoSRgmQxtZ6EW3YC1Z1OB1FKHy3VPHANMSmJqUiTyX1E4W+
SDkbEmVRDkMvuK4yP13ANFgGqKVSbaNO1AvOqqmJSUj9VmQ92Cc7WkhwNNre
rKtiilH1OBlGyo4z95FtRC+kLCa+mAcgNETeqq1ESs7wknlk9H3IXTpnM/Zg
nIUcZNiANPreqq1XKIVHK6wCxshde8DDEoCP84Mqyg1QNNr+KKdssQVwteWB
BB/Y2pDXVjzVlSfpoYWwNLXeqKTL7R5O3UjkNMAsG88Xn87ex5QQ4U8gNFje
qJ1ccXG7MqVRuqL+qpmdqE03vupT5CrwNAdxqKMqlCHwqiN923NnfbvEw4bM
WSvMoCYgNGlsqHjmmRGP3Mk+Id6+onIvN+y3RRlUGokANLVxkKS2VBXwsAsO
90gBS9ysPVfq6iub102ANNoGpqbnXBiQyMV+zFIKxxl7GFGilX8S48CQNHJx
qKoum01H5X1Suuy84TH4vnsiv9yPaYcQNKpaJpFl3a5QeJh5Rh55BqRCkdvs
BT1W32mQNCJHppU9nI/H11WKMx1AmBdainUmt7ZHRPtgNB9HoJF2ncn1E5GO
yWkvI1ElKk8zN0he3atwNAXJpps63kXHElLveowtvCJ1UvOn2NUJbIhwNAhH

pL181iBH5/+13/lmNskKOpFtd0ygBNsgNPFx9o02+QJ1O1P0ZS7IynKPO36p
mphDUw4ANDxHppNVmUXwkzKr+f3RGCbBVAY0t7zjkg6ANNxHopS83v3Qc8pE
NOaq4eifmveYbS9Ot3aANDxHjpU2nSjws7lz7KY8ULSZB7XplFq9BAOANMDe
pI328OKya2ArfTZFd4A5wL34zq/oQK2wNOb8Iu0210XdlRinJpRgUEmx57pC
C78QdHJwNMBHjj02+ArwsfZ9geqMFbm0jUG4wzP0+Q0gNOYUjkS23hhwhdyP
7ybxrewadLIIMpMQ8XgQNOYmkEMm3E9wjn382nifbPifkVSuauXlFGywNA7J
kAx22IvQTAQiyA2gV0Oj0Ftq4S5Hfm0gNCJ5AHj22jjwk6mn2BkJbCAW4LZ7
YQlT1CDwNAd5Io+23/hwvVl8u9EbdRTjWXa9Ya1HTqHgNEC6jp022yrwp5Os
+EFK0FJLKTzd3bAvjpNANDzehDk2+WFwvMAZUYm7NUr0SJZbQFCNLQ9QNAfJ
oDc2/j4Gp2FeCyBF9hKNADWYqsH2WnNgNLVaOio2+NpwhBybYJTomKkZFyCP
0krzuvvANLTJmBPm3vZQXhotVBMl0ITHKZA9LyXVWgAwNDxaOjV22oM7/LeD
20YxuOKEcXqNwo2rOCSgNPpxjFz227hwrsuPVYYKcrQGRDsiURQaaW9gNAhH
5o023nN1EdgVTIjBxMZ19/T2mcGiwUagNEJxqo12nNv1F1tkrw6S3RjFWTo9
tmCf9ieQNMBHvJD+2uBwkp6awzGycsvsOz1YZvGJGWSQNPFxtJE83Ub1Keu9
sgx3VjXIea2HiyeQMRXwNIBxoudm3qvwlS41QATvQCIj6fKSAfpuap5QNAfJ
paX22owQ6RfrUC+8J5RMlukBxJsQ2RpQNPpH2gj23exkQ3gGuoXaNKp0Quyh
n6G97scwNLRxouU23N4Q13BrsUF7iuBzf8Y0YsBePOWgNLVH/qe2nftwp9G9
X52PjUtbdCq8wIuaDsSgNEDepWk3X4IQ6ipt8Zs0DEf7YznRJSkiWwEwNKRH
7qq3nL7wpSKddfseBISVAlxxInUuQzUgNLXexsW32Ci78s71OscJrclH6ko1
8zDL6kYANFpsic023p/WE0/jqNyE+OCIJsH8B/kXLC3ANFhaIxu22a7WKiMu
bm30MOZq+2FTzNw1ilHgNPGmdMq/mHnkSh7lMc+EVdR8VcEPXVOwtg7ANFSq
7NK2/esGiPvpS7ARrQVom07sDuh8mT5ANFjyo962+bTWGnyzRvSQQlSUyfIb
uLK2pNoQNLEUpn32+sCQ9FuirftmCvnasO8uZfzkQnlANEJaJKSm241EbWan
ji4LzJXRCnpDFq4GF0kQNFrkKJy0+967MgN/Rqw7sgvJV6RoslpoymDgNELe
9J1/28K7Je8gMRY4zWIJez89Wb8En+qwNLVskaE2nxPO8hY8pKLdEwGuk2gt
j3v64E4gNHDeiJc+3lJ1L1o4WA9C9kZWVwHxbXwu4FiQNFpJKI183KKGhB0n
1UgmVpKdqZjiaG7DGuaQNELkUOU83NlkSPGnvQiXzRfmVawetVyA3cMwNLRs
4h823ZlkRhXUvaPbqPPUXy1ivz0MOrgQNLhsqpq827oTy9hw7hWw08v4DHCM
s0TvSTUgNMCmKJe3uH11FTYNsgqjcGnw3nKjJpE0VzKgNKQmo6mm1MzwtJV9
1s1IJClGd2SDJdUL8OmQNLWK77y22iDEXHK9MBYbPikcjkXnyi8mYEIgNEJa
CJ/21V/EU9kEtC8SL4p03cLkY1opDl8ANPr+o7dm3uiGuzosN+MdMNaWRrur
llsOgTrgNOLyla+23ARkRjfqXczvfkc9lyYxtzQ67lfwNHAU5kK/3qJWAN4i
J3OJ1nKSzwDCfkSR/wYANEL+lvj+nfqH24rR5FGOGwu4OXXbCsRyPvfQNHD+
rqE22paT1Py2xgspq6bJgAy4yh3LiuqwNLhspJ/22/vkSvrIU5ZL9ZYXVXTu
aH2ZJ1IwNOL+lqw82MTEaSyeuE8BIu53QsKvFq2QUkiQNLj+rJHn2fyRGXOj
GzipEtpjKNtFKyktBvIwNEL6LKc2uuX1Ix2eo67CZ6MqnblMBV3jQoOANFrW
PkNntWn1FRd/z8EBQoLPNQLTFs1F30wQNFpIDKx2mJL1H1W241hkSBmWqJNI
M3RFunfANFrkbJ0+3I3EUHYsg8U1hSQsBT9AQH+DoSygNKRIKKb39hR1FeZK

qCnKGyyMDKLuCeQi6njwNFTkLZH2+EJ1L1JoVrc5q9YE9DqLKU4ahzogNKQ8
qMKn3sbkbokkcD6cUgNUiJ+aSv8izIAANLjcKZV2/xhwqZmMwGGkOcrTWJDI
ti0KP5gQNKS0KqM29yI70rapnokb6uVu4yiN8xO4lvpwNKRsn4s3mS3QZfBn
Eqtl46sVNCBZa42wENywNKTehyem+HZ1JP78lMw4AxLjHyc1JWoCU6zwNHVI
Yi5Wv2dQQlUcsu9L14o3aZxRIqhWjtHQNAjex5+3VOd1CFCkmIvJ6gTk7n6e
Js1FXhtQNHXe7iQ/34xwtqHsGZUIHTI3VJT1bT8xDgbANKRJBym22GX1F4SL
LLN0QTMMcGjJ+8lwgmcQNLjkd5s822Vwxp/uzkeCLVh6gBJMMOKKGNVANLhJ
RqNm+StwiTU5gdKGscmvvVIHcPKx+KlANOIGhTR2neNQenwcvNkpvGYl6TjJ
nb5C0m7wNLWmQ9S23QS7CqFeM1IvlZz+l+g7DRg1LrMgNOLkfh7n+eTwthOr
Tug+NAkVFdMbEfJ3Rz2wNNp5LZ1m2soySj4Mrcbtk2xLhQFu7HxI5gMwNAnk
bqNmu/Cq+qPvwrh2lPcpSmw5dF0pFdCANLiKqJ0uzN31IRJo0dvZUN+bVT9q
gtGQVLgwNKTkcK0+0D/wio1KmJghUdkKwADIvt9mwkJgNFrkKk1nsM9winTT
tHD8bFuBZs0N8v8c3bgwNLXkKqzmm5PwjihtcLwf7nB3KZcde4iW1wbANLRJ
FLK2vHtwiVccrwOWnIZqoIVFB5ULTg1QNEJsxD0u0zh1GtPp/EPX7JesgRyG
sziRVJOgNNxsjQg22yJ1JYd6buLleOIeoYSEOhkFMBtQNAgmqyy2lO/1IJUF
vtFh81H5nMN8G/oih/4gNKRJXR282kJQT0hxODBavFZOpexS2nHu5kLQNNz+
8yM82EX1HDOigtKHUZSxekzwKgLoz5TANNxaLsXX2CRwsHf+al9apSKHPkTP
QPTLAPKwNNwXhry82T51BX7HFRQJOkkt9jV5KUjKrU4gNLhs7swumVNwg1E6
sVWgduCJmikVl82m4MtQNAhsssNuvHl1MT0gw91tsifZX2gvVx6d2CBgNAkX
zJy1nBfwvz9mUWcPtkTDB2k/Tp0GCEdwNKRJU0Em9EFWlAXWfbsI+VkZw7VC
XhUBu9swNKTWDRM2mtm7MSNWUp/CVaWn8fJ13pG3DFPQNHVJKBQz2YF1Hz2z
RqSZkL7WRb9ChmCwguyANLRr8rDWnXZO9mt8bKjYMzCDCXs2kbrT5HmwNOLe
gZp22zwBq75SOwfZ5PMhek97mZrFpGiwNNxIBMH2n6kHwVXWZj4gLuISOLKz
mEtbdkxANOI8gL2z0YQJiOsunjjfMFB4kaewEl44rD/ANLXkchb224FWpTQT
bEhQH8B0Sx5Zb8Q/tfAgNPomgSk2mBAyRukc8oQBKSaYuDYLNBMpp+oANA28
ABG23g3O6lFhFHzXIqBLCQfkvHlRQ0CANOKnNsG00lv1HJKFlZtfIO4zUtxS
nBeSgkkgNLVJBASmvlRD/oMsx4OoInBtCz8Fp8CGpquwNHj0Db3ynEG6MhNL
r9Nq5RXu7twCFNyW4GcgNKJIMte7lYTD0E4G2nR6czB8jJ8Rmy3VBOLANAla
EMN01RjwpxfHR7j5rXRcK8YtLn9LvS0QNA1aBDsmm9RWqK6u2Aekps2XnLog
Ho7ZHYTANAnIN/mS07f1HXRsxQLGAaXobtyGzcBta5NgNHhJLtTkX2sA06gL
Vspzewy4XjYsJXIo1BmQNKRIfuW1q/p1G5BBV8rlsVZUVQ3hLfJ1DqlANLh5
ZMskq33wrbj2ZuK6W1RIH9vqtbN3BgAgNLUx1MjXaPh1bO5zus0EEfgZsg2r
g/tgenkgNKRsi6U6kQL1Ia0LtCBGuflcInvJ2NSsNyuANLRaMl86zcHwrybR
UBtxqQFMFbIBuR4i2jDgNLjeuMEscFyA7AeiwJETnskNkT3Exi6dmjgANNzI
VcMU7pArGGe2PvHxMEEhhH7OChJaK5WgNLh5tAei1vl1Fu5/C9PZR6+TaSCD
tAvcwC6ANMBBg0DkmJn1DnnUAJVbo5tah4QSfXf4EmqwNLVaGFlU16Dwk4Gl
CJiII2IV8baa5PPK8HMgNAdmDiozUaJ1DTeHt29mrKtOkgUThtycq4TQNAV5
bt0qjCb1E0yh+iZARhs+AE5tZrlbetxQNLXJNs+TiToGV5a9m1NA191NRy1B

YXlMvXPwNLXaNOWii0RD2e6L6BCtpIFD8STqzDIhm1eANFp5uMmhT2Z1EvRs
r9F68ZJl+gGMTJ1HmDcQNKR5sMdFT27wueapEu7upHXm2XS8kmXQLYcwNFpJ
ONiRQRp1UDKxupWmoN+TpjmouVsEMH7QNPr0NtsYD+b1BfGG5fCnFiJZvzQq
XuMgZoUANFrENttQRglwqeAXnbfA/8Goh8hsKUtPAxwANFprGtWIB9jwpGGu
cSBs8wK9Ro2flqerESjgNPqANtPKAXHQYSsQRvEyRPsPBJDOeAuLhM5QNPrE
NNWAAy/QSdZzA6Y8scc6mL3lmg49d1zANNxrM0SBIv4X9oMZRU+dKXqu62Rg
OZ/CDZlwNDzKUTeCBaRO8YQIAiQBEYCcs0aE+q8cNj4A

--_----------=_103140636136200
content-disposition: attachment; filename="world.gif"
content-transfer-encoding: base64
content-type: image/gif; name="world.gif"

R0lGODlhpQCWALMAAP///7T////eg/bNtIv2Ys2ki/+cAHus7u4xQSCUUjFa
tIs5KQAAYgAAAAAAAAAAACH+GlNvZnR3YXJlOiBNaWNyb3NvZnQgT2ZmaWNl
ACwAAAAApQCWAAAE/hDISau9OOvNu/9gKI5kaZ5oqq7s1Lzw2850vcZ4rtt8
7wM6gnBILApzv6RyFDM6n0Xcckp1wQCHF3S7jVW/vdhhnNVyz84reK2Ckd/l
BhrdPMrYeJAOTjYPdX5RYllGd3mHF3UEbnyAjo9vgXYNiJUSME8vjTh8kTmd
kpOUlniYUJqdqaqrZaejpF+hhQ2stbamma+wSbJptLbAq712uz+4Z6jByptd
usUzw7nL043NzzTR0sGM1GNyXdcsx3OcrJ/dhM3O4SKKc4vloIHJ091q7O20
2cjc3qH0y+wZMjGQDbd96pKN+/OLWpBZJ3asoYeQXx8uDes5ElTiUayM/q3e
ufKHESQ6HFE6eqQCcqFIhiGboVM1b52HRwV9UPL07aWvdCVn0kzJBKeSfu58
MnQ5S2g8STY74IwKrSFKpU0RAlTWhBFUqhpeDBg7FkkYfVexNgUqs1tNkhz1
NCBLd4CXs2nVrtXK81ZPO3F8ya1L916NHHq7RKLTp99QQTEhShVLuGxOcUn1
/oUHZ3Oai5zNzYp22YJYyoQNt8ir2VFnZHBNghK0r/Ql1JUtg43I1KeYeI0x
ktwqj0hmyRhwFy5sGwXaxKKAeTZulXiqv73jJppbGfdugs+hW78+8uTS6b5s
Ks9d1saviuq4jpbN9dvxXNsts7eM7T16kVuN/veVULHBJ0oOZHGXmmpySeFC
cIkBBNxi1BHozSQvvYDAhgjEsJ+DRQECRHXiUaQFcdjR55Ac+vjGIYfrJdjc
ZFPF8R+AGfkhm44qrthihg28uGGMDDZY432+5dgTijBZ2FhkdAQppHI4IKDS
VNkpZaJxxY1nYZaCTOnIkFbmg6WB5ChZ4SZefgmmHUIOWeWL322HJXTajGNd
m07CVZKUcQY6pJl34jkfej326VeUgjZaZwVn3hhhb3wqSpOkcDYa6IyQRmqo
WpVaygw/mgr66G2PWPYpqKJqxGipcZ6aKnOrYjVIq+ZUpCGsdD6KU2G1gnor
rmy+syuvctKoQ1kJ/gY7KbHXYUodsjB+9ytz0jqrDbSdZZsptTNOVRdr2kbJ
7ai5RDElrNbOKmNg5aaZaKs3HnNsh4CaetOy4/rnba2mhNpnL+7ce++mRqLG
nUL/fnqVwG5Kk2KvmnJKwQsGZGyAgpZ5Ei91nJxzLpQwTVxtxbJirPEeHn/M
sj8QE/hfeH9Q7GjKGuc8IZpaDhvJyBBmhVK++W7qawM5J71yty7PC3TQkCEG
6MFifqCy0jozXa5jT186GjNTF131ZFgr3ZJnSBrrdNckB3jiyaZafHXZK6t5
3puurN012m4fIafY1d5Et912A0Zhz3oD/Y+SAct5c51zD75x4aE1luXL/2zn

KnTB+AKerFSSJ731H/IwBUjm20BmY01UB0526HW3bHi0P6HuYzmyxPD4vrDH
DnWbA9ruFu6Itv43p5GHPnqTtC9VhvARj/Y4A8gj3fvSUM+OLjwTcot5rojN
Z3QDDJRfveTWZ5299rLnoEDiMt8DPCDpiR1D+eZbi376vsO7ZnHca4ACBgi/
mYhIQJebkw7wl78NJE90/MuYhDazJx0M8H0FHN4OKEGf2rxgASBcIAPxF64I
QrBsJnlLtyxIwJidZCUT5EcDQBjC+42QgSUc3AM3NhvmrU6AF7wg13CFpUto
LW8fXMALbshE6u1mh5Oj244O9YIgWnGI9EqiI/5uI7tMKFGL5GviCJEnRRP2
b30xsOIVsSiqJH5RIkDo4qFouEQx41B/KDQj9o44CTWuMYMrCuEbD0iiU3yR
jnW0YwMdWEYdphAXQPSjENloKS3WcIuoqNcMwZhIRZ4Pa1CcXLRYFElJvqAA
qKSkm2iIyBHdRRPDgAEicaBIJ7Yrj44URhUl2cICNCCVLtQIK0OIqjvAMhf0
42AY7WixEYFSj3uMFi8xmEpU+hKQAZkhK2UAIpqdRwACCJgLatnMB4ZSlJfi
5SmtCUxs1meYXyymEVnXAHDac0lWIKfVTHhOL+3yj79k5zUVp81hXsEL/hGF
PcGJnYsts4nNdKbolP7Xo1L2UqDWDGYg4flGV8YRZi9Y6D0Z41BPnkqiOtuf
LnWAUWDujaOXPKgORMpQ45jmoTeM6CWeWUbw4aClA30aDjiKE5rW1KY3FaNO
b0M4GJitFmmkJkYz6c5tOKKjrjTqPRuaHJyS8KTFhGAMdDavXa5zqs+pan22
mFWthpMoyYEoWJ3p1MktDAd29WlABeoGM6jVLyHdqkwDS1MKcsCrTsxH3WAg
I5yUrkbD+eulYCBSKcTAqGg7rFeXCoRnKqgBCQhtpCJFIY36VKtFrSxcHYjT
pUIRB6GNrWxjC9vZ2pa2apKs5Qhb2Jny9q3U2ecdQ9BPGNz2uMg9rol0e/9Z
t4YTEM9daGb1MFy59M64yAVtcpV7NjatDQfOZWhIc/Dcej5XO/vKH1jPSVbt
3vYF2+VuZ3jiM+/+FrXmFW9g94te4XIWmmV8L3zjK+BLZfQg0A3vVitL2fLa
60pzBXAZByxaChN4tiiCwTW1mWAFSze/+m1wWjhrhbl29nrtda+FLyxbBF/1
jR52631lqt8AnMjEJV6vhCkaAxZzt7Yx2CaIY8zgITuYsC+wcZH2ueSwoNiR
Ps4udhMwWvCGt8FFbmuSlYzjYv53xxSNMot5VGUjixe/VoaBknlzF8WiWAdi
JnDakAhiLGPWyuVNgVmKsj9HxFnOc6Zzc2U86Of+tqHNIYoUc8TsXikfDnFm
7q2dSdzVLl9CN3vmmGUY3egCrw9HMSbvViuhaea4YEFxnjKGH+mKZoSam3W2
dFVyYwgqddrHFu7emXLxaljXmdSltkutNX3rKJ9ulCxbCJKvLGrxIoJIzvDO
n1dtWvoap7zMLrShDwFt0xRm2qseWB20jeYZy1oc3VkHZVYMbt0Gzc6Ehve2
81Dq0pym2H+uNgC5p+Bmj5rbqQmLqsFNZXdDMtKSvu+8DcKxEuI734qqSa8V
TmlsmFrgD091xFPk4XqSm6HPLsxkCB5uJ1EQ4dJFFX4Bzh+pkLzF+v4dygUr
b2fT+677enmFu7dWPGf/G+EVX81cdDrwfDsWsB0msnPPfWhhJ4zd7BbwTcus
7pkrPZzcHvpJl0h1ER3G6koP+g1ant4x5sCWRvn61ZlNb7JPppZwZ7rA1750
uUfE7ZqFOzntvh26E/rmTheu3j3JA4XTHdETcS1iB1/d/vg9xHw/NHEXz/hF
On7tlo38RyBaeeoV/uqqEXvWOV95zV+MyIgfET5YK9fSm/4SE9/O6gXORBvq
XfQZAHuTcU/vnML69q+Hfd2bw/tS+D6fe/98vKnS5Nlf4vjIZ2bwgXDnJYPI
+XFtfPTlWviVOHTP2O/qVyut1Om78vpeLqj5WRLGEnI/DKw1ZivDL37Paxax
asVPjp0Mw8n8gwH9GGd23Zd2DgVP/lcFABiA2lcVQ2Vv6neAH9F8dnJH6+dG
tpF69JdjivdQENgpdJSB4JGAGAd+/fGBIJhoGKgsEhiCNXSCpeB1N8BJLviC
JOgcNTiD7NeBIoiDPIh9EQAAOw==

```
--_----------=_103140636136200
content-disposition: attachment; filename="applause.amr"
content-transfer-encoding: base64
content-type: audio/amr; name="applause.amr"
```


IyFBTVIKNGfqHiIP/ms6wozH/P/1GnUaZLAAAFoGf+AANLKCFNen3rDuTi81
/6fmCT2smcvFSQ2SMRcANGfqKBo23rypXCW/8WZDwBsin5eEj83cHWUQNF4e
fPE22yhpUKx//LCWk64JN9tC8aU6if5ANCjGFhY23yD+zJd//faQHLxGtRhc
BS2qeUQgNL9Qad223KjHaTBP8/hDm+0UVRyu84g0UAIwNCYafCI2+hxtQGwn
vnDWZsxo9tRNNkfPSWOANOgeAjE22CFzJdCNnWWG08ziZYYZDGjodhzQNJUa
Izr23QwMQkXe6CF4FKYx9EiPY68FAlpwNBuSZpi23Wk3cNRvmWLqLG5YRlA9
uj0sdXuQNJWiQsE2/9JFsLD2hLFBZWydRaa2LqMYxPQgNL/C5gY23Rn+4+33
/Hx58Bmis8fkLRNXUfggNMVhCFk+/wRSSo3tlyseWdbT0vnmIlhYkkMANHrG
Iw023KYOMKc9fywkOUnJqVncrG4cLinwNBq4Ihl22HlHZs9/vxyfBbA5NOWW
TqUHZhbwNBuSEnS336QQ/Djv/ZLFbgStZu/MrpRHWS9QNBu2WP022MBIx5Tv
/5Qs7a3xOgcja6b5EY6gNFyiAIk+20hZmM39/FJa+O5CUackjC/Vac3QNPO4
Kq022RmX+jzvnIhftcg1ht/Ksd5rRHOANBsMMqs22EtQiswW8qOjnPuzn3cL
IIS11ZNgNPgaMFs2/pd83nFu/8wa3Q5nYy0PVyRfS+WgNCgMMHa23abHTDTv
/mapnwBdNBAjQFrA0axwNBvqE/w23xXHX9Gu/XFT5aX6FLAVYzMOqfKwNKzq
LMM23GVWRxtntSPSqcrSHQqnm4chIfKwNCj4zEQ23DM2kq19nXgIuFEWb61O
4EvZOAVgNPPwBir22vn+lSm//NSLGsRGeavUsMOwZguwNGxjjEi220tO0KF/
/DmD9ZuMrZMNT0CuTNpgNPkeBes22Cs6y7zv/GZX/Uw42VOvCleIz/cANGiS
Vhk23+HQgFHv98MIu+ZF4V8qbvBkbVLQNFOUCRX23whQpqY9sD4YyeKGVRx4
10E/dlkANBpo9Bo23VHpayu//eBtBq8KVCCwL+JUmb6QNBzCqgL2/bnMTYh1
9AHGvrkAqrqHuyxGry6gNMw8P+c22aRQvVh9s29aUQTRnRV0ipYBihIwNEhh
FFL+3PpIyzE/tdGaLPsS6+Zz+BjivTxgNBljyqu+2stQgxu9tHM/oOAZ1ItF
NprS7aJQNBtAiVI+3vBB5tP/vJTU7GkOEuLplobYyX7gNEho61833ahWeYhv
+ducPBiqzGI8E8zelLzwNMY2Ahl22NFzKwl/+r1BHzQ3/BlckH56NmBwNL/q
Wai223LQjhj/9YwzhNttQ5EotsNhYyZgNL/ZXWu22WpQvgv996/dFuHx3KcC
ZvAqQ6UQNKyUTuO32fxmngZrv4nBbcgUaDKMv7Bn5iUQNMZl5AI2/P5O2wZP
/v17jHqSWZ7dpKhqpyzwNIXDCi222W0Q8py/lKcynuAGUQ4JTdoiTO/wNKNl
Bmn32/gMKnt//ywZf1QoOXnzvXYnJ26wNOwyLry223Jm4e693ApIFsk0HATh
AHN2cwoANBthIFV+2akT9oU/twqJkrGQbaXhiBJw/HZgNBvQggW2+WYXx/Z/
/f1RW9WH4PlM734S6f9wNKwgRH12/dpWFcJvnmsOnYJbItwuVm2c42mANByU
Qo72+PYwepGf9dzjwsZXC+dsZiNzDHmANEggOTE23jz0VAZ/9quBN30YX1Ma
q7gbHjBQNPiUVJI+3MGZS8M9ueIBI3ZcXsCsbyIEzt/ANBvQiD832k5HaTD/
/IkhPqsZkXDH01s/FrMwNPPgoIy2355HS30//scM1tuUsJBQ8qyDphHwNEgm
Nby+2T3+j5h9viqqsA4EewbXnje6rxjwNGxlMAw22v5HYfZ//E6lm0c+wS53

QCY9VXuwNBa2HSy+3JxQpxV/9hWXZva2Pm2x8tEmcT8QNMYAaZW23YjHaSb/
/lze7JwdyEvvBoGIUcLwNCRAAEq+3tNOxIl99a5aKpVxNjy6gS8TyVuwNEj4
8ay+2oRA9dN/nAJaR8QPR2Pq1GF0YrzwNOyUJGb+21AMWsL//R8aRWmCfAbk
OIPGeRLANCjqc8s23RK/Nbq9/HhvJO4VbCUcXeo+F4rwNIhhVqk22TaQ4zD+
9yP+i0YUysZEwJVl4GqgNGzqBoD23dswUiy//g22AlYqF2+uVThSIEKwNLeU
flc+3DAOUR2v9NjXxAOkclIHD8pKeRjANKHob+s22eG1Dbp//A1vmLvdnRFI
53MHdpvQNBxAXD2+3FUOy1h3tH6kllOVcCRAJkdE2dbANKYAagI32MlO8Bqv
9GvifoajMuWuH65Y4GMgNGruTdq237rO4xTt/djQmJe7ckHC+bEIsj4gNBtA
OP023x5WKXd83z6VGf6h30sCZZ2k8VzQNGioTAO23tNHETl938TvdLbxeGWQ
Jbc16+pwNIzOHv+22xRQpiQ/Xdsd+nRINIhUjATFwypANIwbBQB+3tQX8x5/
/o8GZYrMsovo/oEYMH8gNBa2RGW32cGT7nu/vLMatyiHmoyxur70zVwANKZl
FED2/DDzIHm9/aPIT++SK2mRYhICfS2ANA75neM22hAT9UG/9IJYUChODVFw
2808tpOANGgAe5K239a782c/+ofuRABb0AUzXZS4RZ9gNIxMCLE22S8k13J9
nhcKQmIebn2cJsTznrUANKwgDgb+2YoGu3+/rbxP8iu5iz99/6omAvbgNKzu
e9022oD1YW79/OqSIsLOCIJaJHNT+kdQNBYyeRX226g6EyDu9JUJyHnRaOLq
FY62ZGDANKa4Cok+1EVERxi/9LLRsCRBJI9ZB5kGPH/gNL/8/wc33rIZQ+gv
/mAZuJ2QVmkYGguzKHQANISMAtl22aBZMYM//kr5zM+futiF+/wYxW0ANCxy
MsF+3nhpiOX9vBqUeusVVwDHBzHmSE7wNKbuOHa+3IsZem4tvRLZlKb4uzBt
YjfDen6QNIySQZ022obmhIHn/YpwUULB2WHjEmtYDNHANLcsehW3/oB1Ked1
/jBhH49yU8/AVGJwgeKQNMqUALA33uBIzF6/t3H8iGA4NSf1ZJd4GBvANLer
phM220Y7CiYv/R7qoA5Lc1hsIP0eshegNCm4Afv22mBO89I93MlfDAKou5FN
boC2Eq6gNLcMNni3+Bx/d3z9trGYZeqMOP3htVb9a5HQNPO4BwK23FgQ56t/
vcGNIT8SUFerGhWh3lrANDTubaM2+X7HQOwtpVnvIqSeFDmhRRCBDhiwNJj5
pAQ33t2FFOD8/CZdJt8ugAbVL7TdjKVgNCz1Qfq23xhzleN//1vtp/KuA8DI
goa9G3vQNJS4Yc722Ql07Pa4v/M4JyZ6o+iFDojaxyfANIz59FW22lJEViy/
X7hO12O54sdQ8hRJvtAwNCgGBt123NBzpSl+n+NYAWBL16vpzAqStf2ANMxM
GUz23tQ76SD5uvFkwMJGUC3yogb/lFAANKYmUEN2+qBEbIav2TabxyM3RYWG
RF5sP67gNClLjGo+31J+0+zv3ugK31dczcvhkxiwo6XANGjOdds2/vzzmYM/
3sQY8h9aIzqCwmTEiYQQNIyMA7623MhOxHRtn8lJ5s0Rtk0YUAwJaQXwNJS2
RPy220kGWa5++FMzOkVLmla85XrEDEwgNBsoUE8/2AFzBncP9f8u54e9vSpI
Q/U4s7igNLkmIj02+LDHAoo3/XKvPp3F1dYgJ15TlaNgNDiMFay226gT1xV+
nWkII3L6T1wgTNpG06YgNPkM+ES2+vP1WqM/t3oCx92thf42zhfbIy+ANBuo
I+C239nEUu/p/oKIfGXCXidN1OhCrLNQNGxj4Nm23EotIMB/tQfPYQJYQJZh
nQxnUcUwNBwMKIQ+2xg760X/5PWscFNCMJvJLNkJ2t7gNFOBRJU22F91Pc3l
4pDTO1s/FmYU4MhTyqlANOwmSTZ22JgX+UB9P5uoA2+OSwsHeg9yOOTQNGhl
yBU3+Aw6pEll/r1Xfo/4X+05+XVZylXgNCLIFgQ23Bos3bVuvddQmN7O127/
yJvD86zwNMZj97S221lHesf/+XAmSdIkvwKjl1hcO3TgNIz6BPl23ZgXz6i/
+i+Caw45kmb37fuwwPmwNGhMfAI2+oHNp5Tl9EGSU3OyjSKE7/RvC/8ANGiM

```
JVS23zgQxpg3/BsSxMal88VhoO08X5zwNA6W2Da2/bJESaT99ekktEn5ViVm
bPt815ywNGcMVey22zvHTqNuvNd0sJmqIoGPa/sknifANKaoR732+QBOxXd5
/1ZR/3sYAXC+OzV4+VNgNKgCU+U22ShWI119/SMS1+e9Nb0cT90Xg+1QNMS4
WY/233R1GuY+lV7OQjEdttctiIZtpVMwNBruUaV22a71feMXWXSqRItCdHbw
BXlqK4ZQNBvCaZM2+7eszsi89LUTf810hAyHPDNsCOXwNOy2Pfs22nh08O4u
nVMepHnq+76mI4791BsANH8AF8m2B+hO8T//IauFMAfAMAwpjwVyCMWA
```


```
--_----------=_103140636136200
content-disposition: attachment; filename="Text0000.txt"
content-length: 12
content-transfer-encoding: binary
content-type: text/plain; name="Text0000.txt"

Hello, world
--_----------=_103140636136200--
```

APPENDIX C

UAProf Schema for MMS Handsets

```
<?xml version="1.0"?>
<rdf:RDF xmlns:rdf="http://www.w3.org/1999/02/22-rdf-syntax-ns#"
     xmlns:rdfs="http://www.w3.org/2000/01/rdf-schema#"
     xmlns:prf="http://www.wapforum.org/profiles/UAPROF/
        ccppschema-20020710#">

     <rdf:Description ID="Component">
          <rdf:type resource="http://www.w3.org/2000/01/
           rdf-schema#Class"/>
          <rdfs:subClassOf rdf:resource="http://www.w3.org/
           2000/01/rdf-schema#Resource"/>
          <rdfs:label>Component</rdfs:label>
          <rdfs:comment>
      A Component within the CC/PP Schema is a class of related
      properties that describe the capabilities and preferences
      information.
   </rdfs:comment>

     </rdf:Description>
     <!-- ***************************************************** -->
     <!-- ***** Properties shared among the components***** -->

     <rdf:Description ID="component">
          <rdf:type resource="http://www.w3.org/2000/01/
           rdf-schema#Property"/>
```

```
        <rdfs:label>component</rdfs:label>
        <rdfs:comment>
    The component attribute links the various components to the
    root node (profile).
  </rdfs:comment>
   </rdf:Description>

   <rdf:Description ID="defaults">
        <rdfs:type rdf:resource="http://www.w3.org/2000/01/
         rdf-schema#Property"/>
        <rdfs:domain rdf:resource="#HardwarePlatform"/>
        <rdfs:domain rdf:resource="#SoftwarePlatform"/>
        <rdfs:domain rdf:resource="#WapCharacteristics"/>
        <rdfs:domain rdf:resource="#BrowserUA"/>
        <rdfs:domain rdf:resource="#NetworkCharacteristics"/>
        <rdfs:domain rdf:resource="#PushCharacteristics"/>
        <rdfs:comment>
    An attribute used to identify the default capabilities.
  </rdfs:comment>
   </rdf:Description>

   <!-- ***************************************************** -->
   <!-- ***** Component Definitions ***** -->

   <rdf:Description ID="HardwarePlatform">
        <rdf:type resource="http://www.w3.org/2000/01/
         rdf-schema#Class"/>
        <rdfs:subClassOf rdf:resource="#Component"/>
        <rdfs:label>Component: HardwarePlatform</rdfs:label>
        <rdfs:comment>
    The HardwarePlatform component contains properties of the
    device's Hardware, such as display size, supported
    character sets, etc.
  </rdfs:comment>
   </rdf:Description>

   <rdf:Description ID="SoftwarePlatform">
        <rdf:type resource="http://www.w3.org/2000/01/
```

```
          rdf-schema#Class"/>
         <rdfs:subClassOf rdf:resource="#Component"/>
         <rdfs:label>Component: SoftwarePlatform</rdfs:label>
         <rdfs:comment>
     The SoftwarePlatform component contains properties of the
     device's application environment, operating system, and
     installed software.
  </rdfs:comment>
    </rdf:Description>

    <rdf:Description ID="BrowserUA">
         <rdf:type resource="http://www.w3.org/2000/01/
          rdf-schema#Class"/>
         <rdfs:subClassOf rdf:resource="#Component"/>
         <rdfs:label>Component: BrowserUA</rdfs:label>
         <rdfs:comment>
     The BrowserUA component contains attributes related to the
     browser user agent running on the device.
  </rdfs:comment>
    </rdf:Description>

    <rdf:Description ID="NetworkCharacteristics">
         <rdf:type resource="http://www.w3.org/2000/01/
          rdf-schema#Class"/>
         <rdfs:subClassOf rdf:resource="#Component"/>
         <rdfs:label>Component:
NetworkCharacteristics</rdfs:label>
         <rdfs:comment>
     The NetworkCharacteristics component contains properties
     describing the network environment including the supported
     bearers.
  </rdfs:comment>
    </rdf:Description>

    <rdf:Description ID="WapCharacteristics">
         <rdf:type resource="http://www.w3.org/2000/01/
          rdf-schema#Class"/>
         <rdfs:subClassOf rdf:resource="#Component"/>
```

```
        <rdfs:label>Component: WapCharacteristics</rdfs:label>
        <rdfs:comment>
    The WapCharacteristics component contains properties of the
    WAP environment supported by the device.
  </rdfs:comment>
   </rdf:Description>

   <rdf:Description ID="PushCharacteristics">
        <rdf:type resource="http://www.w3.org/2000/01/
         rdf-schema#Class"/>
        <rdfs:subClassOf rdf:resource="#Component"/>
        <rdfs:label>Component: PushCharacteristics</rdfs:label>
        <rdfs:comment>
    The PushCharacteristics component contains properties of
    the device's push capabilities, such as supported content
    mime types.
  </rdfs:comment>
   </rdf:Description>

   <!-- **
   ** In the following property definitions, the defined types
   ** are as follows:
   **
   **    Number:     A positive integer
   **                  [0-9]+
   **    Boolean:    A yes or no value
   **                  Yes|No
   **    Literal:    An alphanumeric string
   **                  [A-Za-z0-9/.\-_]+
   **    Dimension:  A pair of numbers
   **                  [0-9]+x[0-9]+
   **
-->
   <!-- **************************************************** -->
   <!-- ***** Component: HardwarePlatform ***** -->

   <rdf:Description ID="BluetoothProfile">
        <rdf:type rdf:resource="http://www.w3.org/2000/01/
```

```
     rdf-schema#Property"/>
    <rdf:type rdf:resource="http://www.w3.org/2000/01/
     rdf-schema#Bag"/>
    <rdfs:domain rdf:resource="#HardwarePlatform"/>
    <rdfs:comment>
    Description: Supported Bluetooth profiles as defined in
    the Bluetooth specification [BLT].
          Type:          Literal (bag)
          Resolution:    Locked
          Examples:      "dialup", "lanAccess"
     </rdfs:comment>
</rdf:Description>

<rdf:Description ID="BitsPerPixel">
    <rdf:type rdf:resource="http://www.w3.org/2000/01/
     rdf-schema#Property"/>
    <rdfs:domain rdf:resource="#HardwarePlatform"/>
    <rdfs:comment>
 Description:  The number of bits of color or grayscale
         information per pixel, related to the number of
         colors or shades of gray the device can display.
 Type:         Number
 Resolution:   Override
 Examples:     "2", "8"
</rdfs:comment>
</rdf:Description>

<rdf:Description ID="ColorCapable">
    <rdf:type rdf:resource="http://www.w3.org/2000/01/
     rdf-schema#Property"/>
    <rdfs:domain rdf:resource="#HardwarePlatform"/>
    <rdfs:comment>
 Description:  Indicates whether the device's display
 supports color.
               "Yes" means color is supported. "No" means
               the display supports only grayscale or black
               and white.
 Type:         Boolean
```

```
   Resolution:   Override
   Examples:     "Yes", "No"
</rdfs:comment>
  </rdf:Description>

  <rdf:Description ID="CPU">
       <rdf:type rdf:resource="http://www.w3.org/2000/01/
        rdf-schema#Property"/>
       <rdfs:domain rdf:resource="#HardwarePlatform"/>
       <rdfs:comment>
   Description:  Name and model number of the device CPU.
   Type:         Literal
   Resolution:   Locked
   Examples:     "Pentium III", "PowerPC 750"
</rdfs:comment>
  </rdf:Description>

  <rdf:Description ID="ImageCapable">
       <rdf:type rdf:resource="http://www.w3.org/2000/01/
        rdf-schema#Property"/>
       <rdfs:domain rdf:resource="#HardwarePlatform"/>
       <rdfs:comment>
   Description:  Indicates whether the device supports the
                 display of images. If the value is "Yes", the
                 property CcppAccept may list the types of
                 images supported.
   Type:         Boolean
   Resolution:   Locked
   Examples:     "Yes", "No"
</rdfs:comment>
</rdf:Description>

  <rdf:Description ID="InputCharSet">
       <rdf:type rdf:resource="http://www.w3.org/2000/01/
        rdf-schema#Property"/>
       <rdf:type rdf:resource="http://www.w3.org/2000/01/
        rdf-schema#Bag"/>
       <rdfs:domain rdf:resource="#HardwarePlatform"/>
```

```
        <rdfs:comment>
    Description:  List of character sets supported by the
                  device for text entry. Property's value is a
                  list of character sets, where each item in
                  the list is a character set name,
                  asregistered with IANA.
    Type:         Literal (bag)
    Resolution:   Append
    Examples:     "US-ASCII", "ISO-8859-1", "Shift_JIS"
</rdfs:comment>
  </rdf:Description>

  <rdf:Description ID="Keyboard">
        <rdf:type rdf:resource="http://www.w3.org/2000/01/
         rdf-schema#Property"/>
        <rdfs:domain rdf:resource="#HardwarePlatform"/>
        <rdfs:comment>
    Description:  Type of keyboard supported by the device, as
                  an indicator of ease of text entry.
    Type:         Literal
    Resolution:   Locked
    Examples:     "Disambiguating", "Qwerty", "PhoneKeypad"
</rdfs:comment>
  </rdf:Description>

  <rdf:Description ID="Model">
        <rdf:type rdf:resource="http://www.w3.org/2000/01/
         rdf-schema#Property"/>
        <rdfs:domain rdf:resource="#HardwarePlatform"/>
        <rdfs:comment>
    Description:  Model number assigned to the terminal device
                  by the vendor or manufacturer.
    Type:         Literal
    Resolution:   Locked
    Examples:     "Mustang GT", "Q30"
</rdfs:comment>
  </rdf:Description>
```

```
  <rdf:Description ID="NumberOfSoftKeys">
      <rdf:type rdf:resource="http://www.w3.org/2000/01/
       rdf-schema#Property"/>
      <rdfs:domain rdf:resource="#HardwarePlatform"/>
      <rdfs:comment>
   Description:  Number of soft keys available on the device.
   Type:         Number
   Resolution:   Locked
   Examples:     "3", "2"
</rdfs:comment>
  </rdf:Description>

  <rdf:Description ID="OutputCharSet">
      <rdf:type rdf:resource="http://www.w3.org/2000/01/
       rdf-schema#Property"/>
      <rdf:type rdf:resource="http://www.w3.org/2000/01/
       rdf-schema#Bag"/>
      <rdfs:domain rdf:resource="#HardwarePlatform"/>
      <rdfs:comment>
   Description:  List of character sets supported by the
                 device for output to the display. Property
                 value is a list of character sets, where each
                 item in the list is a character set name, as
                 registered with IANA.
   Type:         Literal (bag)
   Resolution:   Append
   Examples:     "US-ASCII", "ISO-8859-1", "Shift_JIS"
</rdfs:comment>
  </rdf:Description>

  <rdf:Description ID="PixelAspectRatio">
      <rdf:type rdf:resource="http://www.w3.org/2000/01/
       rdf-schema#Property"/>
      <rdfs:domain rdf:resource="#HardwarePlatform"/>
      <rdfs:comment>
   Description:  Ratio of pixel width to pixel height.
   Type:         Dimension
   Resolution:   Locked
```

```
  Examples:      "1x2"
</rdfs:comment>
 </rdf:Description>

 <rdf:Description ID="PointingResolution">
      <rdf:type rdf:resource="http://www.w3.org/2000/01/
       rdf-schema#Property"/>
      <rdfs:domain rdf:resource="#HardwarePlatform"/>
      <rdfs:comment>
  Description:  Type of resolution of the pointing accessory
                supported by the device.
  Type:         Literal
  Resolution:   Locked
  Examples:     "Character", "Line", "Pixel"
</rdfs:comment>
 </rdf:Description>

 <rdf:Description ID="ScreenSize">
      <rdf:type rdf:resource="http://www.w3.org/2000/01/
       rdf-schema#Property"/>
      <rdfs:domain rdf:resource="#HardwarePlatform"/>
      <rdfs:comment>
  Description:  The size of the device's screen in units of
                pixels, composed of the screen width and the
                screen height.
  Type:         Dimension
  Resolution:   Locked
  Examples:     "160x160", "640x480"
</rdfs:comment>
 </rdf:Description>

 <rdf:Description ID="ScreenSizeChar">
      <rdf:type rdf:resource="http://www.w3.org/2000/01/
       rdf-schema#Property"/>
      <rdfs:domain rdf:resource="#HardwarePlatform"/>
      <rdfs:comment>
  Description:  Size of the device's screen in units of
                characters, composed of the screen width and
```

```
                  screen height. The device's standard font
                  should be used to determine this property's
                  value. (Number of characters per row)x(Number
                  of rows). In calculating this attribute
                  use the largest character in the device's
                  default font.
   Type:          Dimension
   Resolution:    Locked
   Examples:      "12x4", "16x8"
</rdfs:comment>
  </rdf:Description>

  <rdf:Description ID="StandardFontProportional">
       <rdf:type rdf:resource="http://www.w3.org/2000/01/
        rdf-schema#Property"/>
       <rdfs:domain rdf:resource="#HardwarePlatform"/>
       <rdfs:comment>
   Description:   Indicates whether the device's standard font
                  is proportional.
   Type:          Boolean
   Resolution:    Locked
   Examples:      "Yes", "No"
</rdfs:comment>
  </rdf:Description>

  <rdf:Description ID="SoundOutputCapable">
       <rdf:type rdf:resource="http://www.w3.org/2000/01/
        rdf-schema#Property"/>
       <rdfs:domain rdf:resource="#HardwarePlatform "/>
       <rdfs:comment>
   Description:   Indicates whether the device supports sound
                  output through an external speaker, headphone
                  jack, or other sound output mechanism.
   Type:          Boolean
   Resolution:    Locked
   Examples:      "Yes", "No"
</rdfs:comment>
  </rdf:Description>
```

```
  <rdf:Description ID="TextInputCapable">
      <rdf:type rdf:resource="http://www.w3.org/2000/01/
       rdf-schema#Property"/>
      <rdfs:domain rdf:resource="#HardwarePlatform"/>
      <rdfs:comment>
   Description:  Indicates whether the device supports alpha-
                 numeric text entry. "Yes" means the device
                 supports entry of both letters and digits.
                 "No" means the device supports only entry of
                 digits.
   Type:         Boolean
   Resolution:   Locked
   Examples:     "Yes", "No"
</rdfs:comment>
  </rdf:Description>

  <rdf:Description ID="Vendor">
      <rdf:type rdf:resource="http://www.w3.org/2000/01/
       rdf-schema#Property"/>
      <rdfs:domain rdf:resource="#HardwarePlatform"/>
      <rdfs:comment>
   Description:  Name of the vendor manufacturing the terminal
                 device.
   Type:         Literal
   Resolution:   Locked
   Examples:     "Ford", "Lexus"
</rdfs:comment>
  </rdf:Description>

  <rdf:Description ID="VoiceInputCapable">
      <rdf:type rdf:resource="http://www.w3.org/2000/01/
       rdf-schema#Property"/>
      <rdfs:domain rdf:resource="#HardwarePlatform"/>
      <rdfs:comment>
   Description:  Indicates whether the device supports any
                 form of voice input, including speech
                 recognition. This includes voice-enabled
```

```
                  browsers.
   Type:          Boolean
   Resolution:    Locked
   Examples:      "Yes", "No"
 </rdfs:comment>
  </rdf:Description>
  <!--
*************************************************************** -->
  <!-- ***** Component: SoftwarePlatform ***** -->

  <rdf:Description ID="AcceptDownloadableSoftware">
      <rdf:type rdf:resource="http://www.w3.org/2000/01/
       rdf-schema#Property"/>
      <rdfs:domain rdf:resource="#SoftwarePlatform"/>
      <rdfs:comment>
   Description:   Indicates the user's preference on whether to
                  accept downloadable software.
   Type:          Boolean
   Resolution:    Locked
   Examples:      "Yes", "No"
 </rdfs:comment>
  </rdf:Description>

  <rdf:Description ID="AudioInputEncoder">
      <rdf:type rdf:resource="http://www.w3.org/2000/01/
       rdf-schema#Property"/>
      <rdf:type rdf:resource="http://www.w3.org/2000/01/
       rdf-schema#Bag"/>
      <rdfs:domain rdf:resource="#SoftwarePlatform"/>
      <rdfs:comment>
   Description:   List of audio input encoders supported by the
                  device.
   Type:          Literal (bag)
   Resolution:    Append
   Example:       "G.711"
 </rdfs:comment>
  </rdf:Description>
```

```
<rdf:Description ID="CcppAccept">
        <rdf:type rdf:resource="http://www.w3.org/2000/01/
         rdf-schema#Property"/>
        <rdf:type rdf:resource="http://www.w3.org/2000/01/
         rdf-schema#Bag"/>
        <rdfs:domain rdf:resource="#SoftwarePlatform"/>
        <rdfs:comment>
  Description:  List of content types the device supports.
                Property value is a list of MIME types, where
                each item in the list is a content type
                descriptor as specified by RFC 2045.
  Type:         Literal (bag)
  Resolution:   Append
  Examples:     "text/html", "text/plain", "text/html",
                "image/gif"
</rdfs:comment>
 </rdf:Description>

 <rdf:Description ID="CcppAccept-Charset">
        <rdf:type rdf:resource="http://www.w3.org/2000/01/
         rdf-schema#Property"/>
        <rdf:type rdf:resource="http://www.w3.org/2000/01/
         rdf-schema#Bag"/>
        <rdfs:domain rdf:resource="#SoftwarePlatform"/>
        <rdfs:comment>
  Description:  List of character sets the device supports.
                Property value is a list of character sets,
                where each item in the list is a character
                set name registered with IANA.

  Type:         Literal (bag)
  Resolution:   Append
  Examples:     "US-ASCII", "ISO-8859-1", "Shift_JIS"
</rdfs:comment>
 </rdf:Description>

 <rdf:Description ID="CcppAccept-Encoding">
        <rdf:type rdf:resource="http://www.w3.org/2000/01/
```

```
        rdf-schema#Property"/>
        <rdf:type rdf:resource="http://www.w3.org/2000/01/
        rdf-schema#Bag"/>
        <rdfs:domain rdf:resource="#SoftwarePlatform"/>
        <rdfs:comment>
    Description:  List of transfer encodings the device
                  supports. Property value is a list of
                  transfer encodings, where each item in the
                  list is a transfer encoding name as specified
                  by RFC 2045 and registered with IANA.
    Type:         Literal (bag)
    Resolution:   Append
    Examples:     "base64", "quoted-printable"
</rdfs:comment>
   </rdf:Description>

   <rdf:Description ID="CcppAccept-Language">
        <rdf:type rdf:resource="http://www.w3.org/2000/01/
        rdf-schema#Property"/>
        <rdf:type rdf:resource="http://www.w3.org/2000/01/
        rdf-schema#Seq"/>
        <rdfs:domain rdf:resource="#SoftwarePlatform"/>
        <rdfs:comment>
    Description:  List of preferred document languages. If a
                  resource is available in more than one
                  natural language, the server can use this
                  property to determine which version of the
                  resource to send to the device. The first
                  item in the list should be considered the
                  user's first choice, the second the second
                  choice, and so on. Property value is a list
                  of natural languages, where each item in the
                  list is the name of a language as defined by
                  RFC 3066[RFC3066].
    Type:         Literal (sequence)
    Resolution:   Append
    Examples:     "zh-CN", "en", "fr"
 </rdfs:comment>
```

```
</rdf:Description>


<rdf:Description ID="DownloadableSoftwareSupport">
     <rdf:type rdf:resource="http://www.w3.org/2000/01/
      rdf-schema#Property"/>
     <rdf:type rdf:resource="http://www.w3.org/2000/01/
      rdf-schema#Bag"/>
     <rdfs:domain rdf:resource="#SoftwarePlatform"/>
     <rdfs:comment>
 Description:  List of executable content types which the
               device supports and which it is willing to
               accept from the network. The property value
               is a list of MIME types, where each item in
               the list is a content type descriptor as
               specified by RFC 2045.
 Type:         Literal (bag)
 Resolution:   Locked
 Examples:     "application/x-msdos-exe"
</rdfs:comment>
</rdf:Description>

<rdf:Description ID="JavaEnabled">
     <rdf:type rdf:resource="http://www.w3.org/2000/01/
      rdf-schema#Property"/>
     <rdfs:domain rdf:resource="#SoftwarePlatform"/>
     <rdfs:comment>
 Description:  Indicates whether the device supports a Java
               virtual machine.
 Type:         Boolean
 Resolution:   Locked
 Examples:     "Yes", "No"
</rdfs:comment>
</rdf:Description>

<rdf:Description ID="JavaPlatform">
     <rdf:type rdf:resource="http://www.w3.org/2000/01/
      rdf-schema#Property"/>
```

```
        <rdf:type rdf:resource="http://www.w3.org/2000/01/
         rdf-schema#Bag"/>
        <rdfs:domain rdf:resource="#SoftwarePlatform"/>
        <rdfs:comment>
    Description: The list of JAVA platforms and profiles
                 installed in the device. Each item in the list
                 is a name token describing compatibility with
                 the name and version of the java platform
                 specification or the name and version of the
                 profile specification name (if profile is
                 included in the device)

    Type:         Literal (bag)
    Resolution:   Append
    Examples: "Pjava/1.1.3-compatible", "MIDP/1.0-compatible",
              "J2SE/1.0-compatible"
</rdfs:comment>
  </rdf:Description>

  <rdf:Description ID="JVMVersion">
        <rdf:type rdf:resource="http://www.w3.org/2000/01/
         rdf-schema#Property"/>
        <rdf:type rdf:resource="http://www.w3.org/2000/01/
         rdf-schema#Bag"/>
        <rdfs:domain rdf:resource="#SoftwarePlatform"/>
        <rdfs:comment>
    Description:  List of the Java virtual machines installed
                  on the device. Each item in the list is a
                  name token describing the vendor and version
                  of the VM.
    Type:         Literal (bag)
    Resolution:   Append
    Examples:     "SunJRE/1.2", "MSJVM/1.0"
</rdfs:comment>
  </rdf:Description>

  <rdf:Description ID="MexeClassmarks">
        <rdf:type rdf:resource="http://www.w3.org/2000/01/
```

```
            rdf-schema#Property"/>
           <rdf:type rdf:resource="http://www.w3.org/2000/01/
            rdf-schema#Bag"/>
           <rdfs:domain rdf:resource="#SoftwarePlatform "/>
           <rdfs:comment>
Description: List of MExE classmarks supported by the device.
              Value "1" means the MExE device supports WAP, value
              "2" means that MExE device supports Personal Java
              and value "3" means that MExE device supports MIDP
              applications.
           Type:          Literal (bag)
           Resolution:    Locked
           Examples:      "1", "3"
   </rdfs:comment>
     </rdf:Description>

     <rdf:Description ID="MexeSpec">
           <rdf:type rdf:resource="http://www.w3.org/2000/01/
            rdf-schema#Property"/>
           <rdfs:domain rdf:resource="#SoftwarePlatform"/>
           <rdfs:comment>
      Description:  Class mark specialization. Refers to the
                    first two digits of the version of the MExE
                    Stage 2 spec.
      Type:         Literal
      Resolution:   Locked
      Examples:     "7.02"
   </rdfs:comment>
     </rdf:Description>

     <rdf:Description ID="MexeSecureDomains">
           <rdf:type rdf:resource="http://www.w3.org/2000/01/
            rdf-schema#Property"/>
           <rdfs:domain rdf:resource="#SoftwarePlatform"/>
           <rdfs:comment>
            Description: Indicates whether the device's supports
                          MExE security domains. "Yes" means that
                          security domains are supported in
```

```
                     accordance with MExE specifications
                     identified by the MexeSpec attribute. "No"
                     means that security domains are not
                     supported and the device has only
                     untrusted domain (area).
   Type:          Boolean
   Resolution:    Locked
   Examples:      "Yes", "No"
</rdfs:comment>
  </rdf:Description>

  <rdf:Description ID="OSName">
       <rdf:type rdf:resource="http://www.w3.org/2000/01/
        rdf-schema#Property"/>
       <rdfs:domain rdf:resource="#SoftwarePlatform"/>
       <rdfs:comment>
   Description:  Name of the device's operating system.
   Type:          Literal
   Resolution:    Locked
   Examples:      "Mac OS", "Windows NT"
</rdfs:comment>
  </rdf:Description>
  <rdf:Description ID="OSVendor">
       <rdf:type rdf:resource="http://www.w3.org/2000/01/
        rdf-schema#Property"/>
       <rdfs:domain rdf:resource="#SoftwarePlatform"/>
       <rdfs:comment>
   Description:  Vendor of the device's operating system.
   Type:          Literal
   Resolution:    Locked
   Examples:      "Apple", "Microsoft"
</rdfs:comment>
  </rdf:Description>

  <rdf:Description ID="OSVersion">
       <rdf:type rdf:resource="http://www.w3.org/2000/01/
        rdf-schema#Property"/>
       <rdfs:domain rdf:resource="#SoftwarePlatform"/>
```

```
        <rdfs:comment>
    Description:  Version of the device's operating system.
    Type:         Literal
    Resolution:   Locked
    Examples:     "6.0", "4.5"
</rdfs:comment>
  </rdf:Description>
  <rdf:Description ID="RecipientAppAgent">
        <rdf:type rdf:resource="http://www.w3.org/2000/01/
         rdf-schema#Property"/>
        <rdfs:domain rdf:resource="#SoftwarePlatform"/>
        <rdfs:comment>
    Description:  User agent associated with the current
                  request. Value should match the name of one
                  of the components in the profile. A component
                  name is specified by the ID attribute on the
                  prf:Component element containing the
                  properties of that component.
    Type:         Literal
    Resolution:   Locked
    Examples:     "BrowserMail"
</rdfs:comment>
  </rdf:Description>

  <rdf:Description ID="SoftwareNumber">
        <rdf:type rdf:resource="http://www.w3.org/2000/01/
         rdf-schema#Property"/>
        <rdfs:domain rdf:resource="#SoftwarePlatform"/>
        <rdfs:comment>
    Description:  Version of the device-specific software
                  (firmware) to which the device's low-level
                  software conforms.
    Type:         Literal
    Resolution:   Locked
    Examples:     "2"
</rdfs:comment>
  </rdf:Description>
```

```
<rdf:Description ID="VideoInputEncoder">
    <rdf:type rdf:resource="http://www.w3.org/2000/01/
     rdf-schema#Property"/>
    <rdf:type rdf:resource="http://www.w3.org/2000/01/
     rdf-schema#Bag"/>
    <rdfs:domain rdf:resource="#SoftwarePlatform"/>
    <rdfs:comment>
 Description:  List of video input encoders supported by the
               device.
 Type:         Literal (bag)
 Resolution:   Append
 Examples:     "MPEG-1", "MPEG-2", "H.261"
</rdfs:comment>
 </rdf:Description>

<rdf:Description ID="Email-URI-Schemes">
    <rdf:type rdf:resource="http://www.w3.org/2000/01/
     rdf-schema#Property"/>
    <rdf:type rdf:resource="http://www.w3.org/2000/01/
     rdf-schema#Bag"/>
    <rdfs:domain rdf:resource="#SoftwarePlatform"/>
    <rdfs:comment>
      Description:  List of URI schemes the device supports
                    for accessing e-mail. Property value is
                    a list of URI schemes, where each item
                    in the list is a URI scheme as defined
                    in RFC 2396.
      Type:         Literal (bag)
      Resolution:   Override
      Examples:     "pop", "imap", "http", "https"
</rdfs:comment>
    </rdf:Description>

<!-- ************************************************ -->
<!-- ***** Component: NetworkCharacteristics ***** -->

<rdf:Description ID="SupportedBluetoothVersion">
    <rdf:type rdf:resource="http://www.w3.org/2000/01/
```

```
            rdf-schema#Property"/>
           <rdfs:domain rdf:resource="#NetworkCharacteristics "/>
           <rdfs:comment>
           Description: Supported Bluetooth version.
                  Type:           Literal
                  Resolution:     Locked
                  Examples:       "1.0"
            </rdfs:comment>
      </rdf:Description>

      <rdf:Description ID="CurrentBearerService">
           <rdf:type rdf:resource="http://www.w3.org/2000/01/
            rdf-schema#Property"/>
           <rdfs:domain rdf:resource="#NetworkCharacteristics"/>
           <rdfs:comment>
       Description:  The bearer on which the current session was
                     opened.
       Type:         Literal
       Resolution:   Locked
       Examples:     "OneWaySMS", "GUTS", "TwoWayPacket"
    </rdfs:comment>
      </rdf:Description>

      <rdf:Description ID="SecuritySupport">
           <rdf:type rdf:resource="http://www.w3.org/2000/01/
            rdf-schema#Property"/>
           <rdf:type rdf:resource="http://www.w3.org/2000/01/
            rdf-schema#Bag"/>
           <rdfs:domain rdf:resource="#NetworkCharacteristics"/>
           <rdfs:comment>
       Description:  List of types of security or encryption
                     mechanisms supported by the device.
       Type:         Literal (bag)
       Resolution:   Locked
       Example:      "WTLS-1", WTLS-2", "WTLS-3", "signText",
"PPTP"
    </rdfs:comment>
      </rdf:Description>
```

```
<rdf:Description ID="SupportedBearers">
     <rdf:type rdf:resource="http://www.w3.org/2000/01/
      rdf-schema#Property"/>
     <rdf:type rdf:resource="http://www.w3.org/2000/01/
      rdf-schema#Bag"/>
     <rdfs:domain rdf:resource="#NetworkCharacteristics"/>
     <rdfs:comment>
 Description:  List of bearers supported by the device.
 Type:         Literal (bag)
 Resolution:   Locked
 Examples:     "GPRS", "GUTS", "SMS", CSD", "USSD"
</rdfs:comment>
 </rdf:Description>
 <!--
*********************************************************** -->
 <!-- ***** Component: BrowserUA ***** -->

 <rdf:Description ID="BrowserName">
     <rdf:type rdf:resource="http://www.w3.org/2000/01/
      rdf-schema#Property"/>
     <rdfs:domain rdf:resource="#BrowserUA"/>
     <rdfs:comment>
 Description:  Name of the browser user agent associated
               with the current request.
 Type:         Literal
 Resolution:   Locked
 Examples:     "Mozilla", "MSIE", "WAP42"
</rdfs:comment>
 </rdf:Description>

 <rdf:Description ID="BrowserVersion">
     <rdf:type rdf:resource="http://www.w3.org/2000/01/
      rdf-schema#Property"/>
     <rdfs:domain rdf:resource="#BrowserUA"/>
     <rdfs:comment>
 Description:  Version of the browser.
 Type:         Literal
```

```
    Resolution:   Locked
    Examples:     "1.0"
</rdfs:comment>
  </rdf:Description>

  <rdf:Description ID="DownloadableBrowserApps">
       <rdf:type rdf:resource="http://www.w3.org/2000/01/
        rdf-schema#Property"/>
       <rdf:type rdf:resource="http://www.w3.org/2000/01/
        rdf-schema#Bag"/>
       <rdfs:domain rdf:resource="#BrowserUA"/>
       <rdfs:comment>
    Description:  List of executable content types which the
                  browser supports and which it is willing to
                  accept from the network. The property value
                  is a list of MIME types,
                  where each item in the list is a content type
                  descriptor as specified by RFC 2045.
    Type:         Literal (bag)
    Resolution:   Append
    Examples:     "application/x-java-vm/java-applet"
</rdfs:comment>
  </rdf:Description>

  <rdf:Description ID="FramesCapable">
       <rdf:type rdf:resource="http://www.w3.org/2000/01/
        rdf-schema#Property"/>
       <rdfs:domain rdf:resource="#BrowserUA"/>
       <rdfs:comment>
    Description:  Indicates whether the browser is capable of
                  displaying frames.
    Type:         Boolean
    Resolution:   Override
    Examples:     "Yes", "No"
</rdfs:comment>
  </rdf:Description>

  <rdf:Description ID="HtmlVersion">
```

```
        <rdf:type rdf:resource="http://www.w3.org/2000/01/
         rdf-schema#Property"/>
        <rdfs:domain rdf:resource="#BrowserUA"/>
        <rdfs:comment>
    Description:  Version of HyperText Markup Language (HTML)
                  supported by the browser.
    Type:         Literal
    Resolution:   Locked
    Examples:     "2.0", "3.2", "4.0"
</rdfs:comment>
  </rdf:Description>

  <rdf:Description ID="JavaAppletEnabled">
        <rdf:type rdf:resource="http://www.w3.org/2000/01/
         rdf-schema#Property"/>
        <rdfs:domain rdf:resource="#BrowserUA"/>
        <rdfs:comment>
    Description:  Indicates whether the browser supports Java
                  applets.
    Type:         Boolean
    Resolution:   Locked
    Examples:     "Yes", "No"
</rdfs:comment>
  </rdf:Description>

  <rdf:Description ID="JavaScriptEnabled">
        <rdf:type rdf:resource="http://www.w3.org/2000/01/
         rdf-schema#Property"/>
        <rdfs:domain rdf:resource="#BrowserUA"/>
        <rdfs:comment>
    Description:  Indicates whether the browser supports
                  JavaScript.
    Type:         Boolean
    Resolution:   Locked
    Examples:     "Yes", "No"
</rdfs:comment>
  </rdf:Description>
```

```
    <rdf:Description ID="JavaScriptVersion">
        <rdf:type rdf:resource="http://www.w3.org/2000/01/
         rdf-schema#Property"/>
        <rdfs:domain rdf:resource="#BrowserUA"/>
        <rdfs:comment>
     Description:  Version of the JavaScript language supported
                   by the browser.
     Type:         Literal
     Resolution:   Locked
     Examples:     "1.4"
  </rdfs:comment>
    </rdf:Description>

    <rdf:Description ID="PreferenceForFrames">
        <rdf:type rdf:resource="http://www.w3.org/2000/01/
         rdf-schema#Property"/>
        <rdfs:domain rdf:resource="#BrowserUA"/>
        <rdfs:comment>
     Description:  Indicates the user's preference for receiving
                   HTML content that contains frames.
     Type:         Boolean
     Resolution:   Locked
     Examples:     "Yes", "No"
  </rdfs:comment>
    </rdf:Description>

    <rdf:Description ID="TablesCapable">
        <rdf:type rdf:resource="http://www.w3.org/2000/01/
         rdf-schema#Property"/>
        <rdfs:domain rdf:resource="#BrowserUA"/>
        <rdfs:comment>
     Description:  Indicates whether the browser is capable of
                   displaying tables.
     Type:         Boolean
     Resolution:   Locked
     Examples:     "Yes", "No"
  </rdfs:comment>
    </rdf:Description>
```

```
  <rdf:Description ID="XhtmlVersion">
      <rdf:type rdf:resource="http://www.w3.org/2000/01/
       rdf-schema#Property"/>
      <rdfs:domain rdf:resource="#BrowserUA"/>
      <rdfs:comment>
   Description:  Version of XHTML supported by the browser.
   Type:         Literal
   Resolution:   Locked
   Examples:     "1.0"
</rdfs:comment>
  </rdf:Description>

  <rdf:Description ID="XhtmlModules">
      <rdf:type rdf:resource="http://www.w3.org/2000/01/
       rdf-schema#Property"/>
      <rdf:type rdf:resource="http://www.w3.org/2000/01/
       rdf-schema#Bag"/>
      <rdfs:domain rdf:resource="#BrowserUA"/>
      <rdfs:comment>
   Description:  List of XHTML modules supported by the
                 browser. Property value is a list of module
                 names, where each item in the list is the
                 name of an XHTML module as defined by the
                 W3C document "Modularization of XHTML",
                 Section 4. List items are separated by white
                 space. Note that the referenced document is a
                 work in progress. Any subsequent changes to
                 the module naming conventions should be
                 reflected in the values of this property.
   Type:         Literal (bag)
   Resolution:   Append
   Examples:     "XHTML1-struct", "XHTML1-blkstruct",
                 "XHTML1-frames"
</rdfs:comment>
  </rdf:Description>

  <!-- ****************************************************** -->
```

```
<!-- ***** Component: WapCharacteristics ***** -->

<rdf:Description ID="SupportedPictogramSet">
     <rdf:type rdf:resource="http://www.w3.org/2000/01/
      rdf-schema#Property"/>
     <rdf:type rdf:resource="http://www.w3.org/2000/01/
      rdf-schema#Bag"/>
     <rdfs:domain rdf:resource="#WapCharacteristics"/>
     <rdfs:comment>
Description: Pictogram classes supported by the device as defined
             in "WAP Pictogram specification".
     Type:          Literal (bag)
     Resolution:    Append
     Examples:      "core", "core/operation", "human"
  </rdfs:comment>
</rdf:Description>

<rdf:Description ID="WapDeviceClass">
     <rdf:type rdf:resource="http://www.w3.org/2000/01/
      rdf-schema#Property"/>
     <rdfs:domain rdf:resource="#WapCharacteristics"/>
     <rdfs:comment>
  Description:  Classification of the device based on
                capabilities as identified in the WAP 1.1
                specifications. Current
                values are "A", "B" and "C".
  Type:         Literal
  Resolution:   Locked
  Examples:     "A"
 </rdfs:comment>
</rdf:Description>

<rdf:Description ID="WapVersion">
     <rdf:type rdf:resource="http://www.w3.org/2000/01/
      rdf-schema#Property"/>
     <rdfs:domain rdf:resource="#WapCharacteristics"/>
     <rdfs:comment>
  Description:  Version of WAP supported.
```

```
   Type:          Literal
   Resolution:    Locked
   Examples:      "1.1", "1.2.1", "2.0"
</rdfs:comment>
  </rdf:Description>

  <rdf:Description ID="WmlDeckSize">
       <rdf:type rdf:resource="http://www.w3.org/2000/01/
        rdf-schema#Property"/>
       <rdfs:domain rdf:resource="#WapCharacteristics"/>
       <rdfs:comment>
   Description:   Maximum size of a WML deck that can be
                  downloaded to the device. This may be an
                  estimate of the maximum size if the true
                  maximum size is not known. Value is number
                  of bytes.
   Type:          Number
   Resolution:    Locked
   Examples:      "4096"
</rdfs:comment>
  </rdf:Description>

  <rdf:Description ID="WmlScriptLibraries">
       <rdf:type rdf:resource="http://www.w3.org/2000/01/
        rdf-schema#Property"/>
       <rdf:type rdf:resource="http://www.w3.org/2000/01/
        rdf-schema#Bag"/>
       <rdfs:domain rdf:resource="#WapCharacteristics"/>
       <rdfs:comment>
   Description:   List of mandatory and optional libraries
                  supported in the device's WMLScript VM.
   Type:          Literal (bag)
   Resolution:    Locked
   Examples:      "Lang", "Float", "String", "URL",
                  "WMLBrowser", "Dialogs", "PSTOR"
</rdfs:comment>
  </rdf:Description>
```

```
<rdf:Description ID="WmlScriptVersion">
     <rdf:type rdf:resource="http://www.w3.org/2000/01/
      rdf-schema#Property"/>
     <rdf:type rdf:resource="http://www.w3.org/2000/01/
      rdf-schema#Bag"/>
     <rdfs:domain rdf:resource="#WapCharacteristics"/>
     <rdfs:comment>
  Description:  List of WMLScript versions supported by the
                device. Property value is a list of version
                numbers, where each item in the list is a
                version string conforming to Version.
  Type:         Literal (bag)
  Resolution:   Append
  Examples:     "1.1", "1.2"
</rdfs:comment>
  </rdf:Description>

 <rdf:Description ID="WmlVersion">
     <rdf:type rdf:resource="http://www.w3.org/2000/01/
      rdf-schema#Property"/>
     <rdf:type rdf:resource="http://www.w3.org/2000/01/
      rdf-schema#Bag"/>
     <rdfs:domain rdf:resource="#WapCharacteristics"/>
     <rdfs:comment>
  Description:  List of WML language versions supported by
                the device. Property value is a list of
                version numbers, where each item in the list
                is a version string conforming
                to Version.
  Type:         Literal (bag)
  Resolution:   Append
  Examples:     "1.1", "2.0"
</rdfs:comment>
  </rdf:Description>

 <rdf:Description ID="WtaiLibraries">
     <rdf:type rdf:resource="http://www.w3.org/2000/01/
      rdf-schema#Property"/>
```

```
        <rdf:type rdf:resource="http://www.w3.org/2000/01/
         rdf-schema#Bag"/>
        <rdfs:domain rdf:resource="#WapCharacteristics"/>
        <rdfs:comment>
    Description:  List of WTAI network common and network
                  specific libraries supported by the device.
                  Property value is a list of WTA library
                  names, where each item in the list list is a
                  library name as specified by "WAP WTAI" and
                  its addendums. Any future addendums to "WAP
                  WTAI" should be reflected in the values of
                  this property.
    Type:         Literal (bag)
    Resolution:   Locked
    Examples:     "WTAVoiceCall", "WTANetText", "WTAPhoneBook",
                  "WTACallLog", "WTAMisc", "WTAGSM",
                  "WTAIS136", "WTAPDC"
</rdfs:comment>
  </rdf:Description>

  <rdf:Description ID="WtaVersion">
        <rdf:type rdf:resource="http://www.w3.org/2000/01/
         rdf-schema#Property"/>
        <rdfs:domain rdf:resource="#WapCharacteristics"/>
        <rdfs:comment>
    Description:  Version of WTA user agent.
    Type:         Literal
    Resolution:   Locked
    Examples:     "1.1"
</rdfs:comment>
  </rdf:Description>

  <rdf:Description ID="DrmClass">
        <rdf:type rdf:resource="http://www.w3.org/2000/01/
         rdf-schema#Property"/>
        <rdfs:domain rdf:resource="#WapCharacteristics"/>
        <rdfs:comment>
    Description:  DRM Conformance Class as defined in OMA-
```

```
                  Download-DRM-v1_0
     Type:        Literal (bag)
     Resolution:  Locked
     Examples:    "ForwardLock", "CombinedDelivery",
                  "SeparateDelivery"
</rdfs:comment>
  </rdf:Description>

  <rdf:Description ID="DrmConstraints">
      <rdf:type rdf:resource="http://www.w3.org/2000/01/
       rdf-schema#Property"/>
      <rdfs:domain rdf:resource="#WapCharacteristics"/>
      <rdfs:comment>
     Description:  DRM permission constraints as defined in OMA-
                   Download-DRMREL-v1_0.  The datetime and
                   interval constraints depend on having a
                   secure clock in the terminal.
     Type:         Literal (bag)
     Resolution:   Locked
     Examples:     "datetime", "interval"
</rdfs:comment>
  </rdf:Description>

  <rdf:Description ID="OmaDownload">
      <rdf:type rdf:resource="http://www.w3.org/2000/01/
       rdf-schema#Property"/>
      <rdfs:domain rdf:resource="#WapCharacteristics"/>
      <rdfs:comment>
     Description:  Supports OMA Download as defined in OMA-
                   Download-OTA-v1_0
     Type:         Boolean
     Resolution:   Locked
     Examples:     "Yes", "No"
</rdfs:comment>
  </rdf:Description>

  <!--
**************************************************************** -->
```

```
<!-- ***** Component: PushCharacteristics ***** -->

<rdf:Description ID="Push-Accept">
    <rdf:type rdf:resource="http://www.w3.org/2000/01/
     rdf-schema#Property"/>
    <rdf:type rdf:resource="http://www.w3.org/2000/01/
     rdf-schema#Bag"/>
    <rdfs:domain rdf:resource="#PushCharacteristics"/>
    <rdfs:comment>
  Description:  List of content types the device supports,
                which can be carried inside the message/http
                entity body when OTA-HTTP is used. Property
                value is a list of MIME types, where each
                item in the list is a content type descriptor
                as specified by RFC 2045.
  Type:         Literal (bag)
  Resolution:   Override
  Examples:     "text/html", "text/plain", "image/gif"
</rdfs:comment>
</rdf:Description>


   <rdf:Description ID="Push-Accept-Charset">
      <rdf:type rdf:resource="http://www.w3.org/2000/01/
       rdf-schema#Property"/>
      <rdf:type rdf:resource="http://www.w3.org/2000/01/
       rdf-schema#Bag"/>
      <rdfs:domain rdf:resource="#PushCharacteristics"/>
      <rdfs:comment>
         Description:  List of character sets the device
                       supports. Property value is a list of
                       character sets, where each item in
                       the list is a character set name
                       registered with IANA.
         Type:         Literal (bag)
         Resolution:   Override
         Examples:     "US-ASCII", "ISO-8859-1", "Shift_JIS"
      </rdfs:comment>
```

```
</rdf:Description>

<rdf:Description ID="Push-Accept-Encoding">
   <rdf:type rdf:resource="http://www.w3.org/2000/01/
    rdf-schema#Property"/>
   <rdf:type rdf:resource="http://www.w3.org/2000/01/
    rdf-schema#Bag"/>
   <rdfs:domain rdf:resource="#PushCharacteristics"/>
   <rdfs:comment>
      Description:  List of transfer encodings the device
                    supports. Property value is a list of
                    transfer encodings, where each item in
                    the list is a transfer encoding name
                    as specified by RFC 2045 and
                    registered with IANA.
      Type:         Literal (bag)
      Resolution:   Override
      Examples:     "base64", "quoted-printable"
  </rdfs:comment>
</rdf:Description>

<rdf:Description ID="Push-Accept-Language">
   <rdf:type rdf:resource="http://www.w3.org/2000/01/
    rdf-schema#Property"/>
   <rdf:type rdf:resource="http://www.w3.org/2000/01/
    rdf-schema#Seq"/>
   <rdfs:domain rdf:resource="#PushCharacteristics"/>
   <rdfs:comment>
      Description:  List of preferred document languages.
                    If a resource is available in more
                    than one natural language, the server
                    can use this property to determine
                    which version of the resource to send
                    to the device. The first item in the
                    list should be considered the user's
                    first choice, thesecond the second
                    choice, and so on. Property value is
                    a list of natural languages, where
```

```
                            each item in the list is the name of a
                            language as defined by RFC
                            3066[RFC3066].
               Type:        Literal (sequence)
               Resolution:  Override
               Examples:    "zh-CN", "en", "fr"
            </rdfs:comment>
    </rdf:Description>

     <rdf:Description ID="Push-Accept-AppID">
          <rdf:type rdf:resource="http://www.w3.org/2000/01/
           rdf-schema#Property"/>
          <rdf:type rdf:resource="http://www.w3.org/2000/01/
           rdf-schema#Bag"/>
          <rdfs:domain rdf:resource="#PushCharacteristics"/>
          <rdfs:comment>
      Description:  List of applications the device supports,
                    where each item in the list is an
                    application-id on absoluteURI format as
                    specified in [PushMsg]. A wildcard ("*") may
                    be used to indicate support for any
                    application.
      Type:         Literal (bag)
      Resolution:   Override
      Examples:     "x-wap-application:wml.ua", "*"
   </rdfs:comment>
     </rdf:Description>

     <rdf:Description ID="Push-MsgSize">
          <rdf:type rdf:resource="http://www.w3.org/2000/01/
           rdf-schema#Property"/>
          <rdfs:domain rdf:resource="#PushCharacteristics"/>
          <rdfs:comment>
      Description:  Maximum size of a push message that the
                    device can handle. Value is number of bytes.
      Type:         Number
      Resolution:   Override
      Examples:     "1024", "1400"
```

```
        </rdfs:comment>
          </rdf:Description>

          <rdf:Description ID="Push-MaxPushReq">
               <rdf:type rdf:resource="http://www.w3.org/2000/01/
                rdf-schema#Property"/>
               <rdfs:domain rdf:resource="#PushCharacteristics"/>
               <rdfs:comment>
            Description:  Maximum number of outstanding push requests
                          that the device
                          can handle.
            Type:         Number
            Resolution:   Override
            Examples:     "1", "5"
        </rdfs:comment>
          </rdf:Description>
     </rdf:RDF>
```

INDEX

Note: Boldface numbers indicate illustrations.

3GPP standard, 20, 207–212
  adaptive multirate (AMR) in, 108
  code division multiple access (CDMA) and, 73
  data calling tone (DCT) in, 95
  media formats and, 90
  roaming and, 200–201
  SMIL and, 72–73
  version number of, 43
3GPP2 standard, 207–212
acknowledgment, 64
adaptive multirate (AMR), 81, 94, 108
  real time transport protocol (RTP) and, 152
  standards for, 211
adaptive multirate wideband (AMR–WB), real time transport protocol (RTP) and, 151–153, **152**
addresses, 43–44
  recipient, 43–44
  sender, 44
Adobe GoLive, 80, 81
Adobe Illustrator, 116
advantages of MMS over WAP, 14
advertising in MMS, 204–205
alpha testing, 122
Alphaworks, 116
AMRConverter, 109
<anchor> tag, 104–106
animated comics service, 9–10
animation, 117–118
application program interfaces (APIs), 179
applications for MMS (*See also* business case for MMS), 2–3, 189
  development of, 15–16
  development tools for, 71–88
  marketing potential for, 185–186
  platforms for, 3–6, **4**
  resources essential to, 16–17
architecture of MMS, 19–33, **21**
  communication bearers and protocols in, 30–32
  distribution of MMS messages and, 20
  general packet radio service (GPRS) connection in, 31–32
  HTTP and, 30
  interfaces in, 23–25, **24**
  Internet protocol (IP) and, 30
  media elements in, 25–26
  message assembly in, 26–28
  message passing in, 28–30, **29**
  MIME in, 27–29, **28, 29**
  MM1 interface in (*See also* message structure and MM1), 23, 29–30, **29, 30**, 31, 35–69, 81, 164, 213–223
  MM3 interface in (*See also* MM3 interface), 23, 24, 30, 147–161, **149**
  MM4 envelopes in, 23
  MM5 interface in, 23
  MM6 interface in, 23
  MM7 interface in (*See also* MM7 interface), 23, 24, 30, 148, 163–175
  MM9 interface in, 148

architecture of MMS *(continued)*
MMSC connection to mobile handset in, 21–23, **21**
MMSC connections to other computers in, 23–25, **21**
model of operation for, 32–33, **32**
multimedia messaging service center (MMSC) in, 21
network nodes in, 31, **31**
receiver in, 20
relay/server (proxy–relay) in, 20–21
router in, 20
sender in, 20
size of messages in, 20, 32–33
SMIL in, 26–28, **27**
standards for (3GPP), 20
support for media elements in, 25
transmission control protocol (TCP) and, 30
user agents in, 22
wireless application protocol (WAP) gateway in, 31
wireless session protocol (WSP) and, 29, 31
<area> tag, 104, 106–107
ASCII files, 15, 38–39, 83, 104
binary MMS representation vs., 126–127
assembly of messages, 26–28
Association SMIL, 73
audio codec selection, 95, **95**
audio content, 25
audio files, 25, 108–114
adaptive multirate (AMR) in, 108
converting files to AMR, 108–109, **110**
GSM and, 108, 111
iMelody in, 112–113
MIDI and SP–MIDI in, 15, 94, 113–114, 212
MPEG files, 111
QCELP and, 111
ringing tones text transfer language (RTTTL) format in, 113
ringtones and, 113
sampling/frequency response in AMR and, 108, **109**
size of, 111, **111**
audio files *(continued)*
speech codecs and, 108
synthetic audio in, 111
WAV files, 111
automatic MMS production, SMIL templates and, 153–157, **155**

Basic SMIL, 73
Beatnik, 116
Beep Science, 67
begin attribute, SMIL and, 77
beta testing, 122
binary files
ASCII MMS representation vs., 126–127
Hello World example, 127–129, 131–136
BitFlash, 117
blocking messages, MM1 interface and, 59
Boston Bruins sports highlights example, 156
BREW initiative, 185, 191–192
Bullet Proof Software, 111
business case for MMS (*See also* applications for MMS), 177–193
application developers' perspective in, 184–185
BREW initiative in, 185, 191–192
distribution paths in, 189–192
getting applications to market and, 185–186
income potential in, 178, 179, **181–182**, 196
market predictions for, 178
marketing potential in, 185–186
opportunities and threats in, **181–182**
participants in, 180, **180**
pricing fundamentals for, 182–184
pricing vs. end user value in, 186–189, **188–189**
Singapore initiative in, 189–191
standards and, 178
strategic view of value chain participants in, 180–182
value chain in, 179–189, **180**

calendars (*See also* vCalendar), 118–119
cartoons (*See* animated comics services)

character sets, 107, **107**
CMG Wireless Data Solutions, 199
Coca–Cola, 204–205
code division multiple access (CDMA), 73
  BREW initiative, 191–192
Cold Fusion, 16
communication bearers, 30–32
composing MMS applications, 79–81, **80**
  testing and, 123–124
  Web–based MMS composers and, 123–124
composite capability/preference profiles (CC/PP), 96, 102
Computer Management Group, 73
computers, multimedia messaging service center (MMSC) connection to, 23–25, **21**
Comverse, 73, 185, 199
Conformance SMIL, 73
construction of MMS message, 78–84
content creation, 5, 148
  application development and, 86
  digital rights management and, 196–198
  media elements of, 25–26
  protection of, 63
  transcoding and content adaptation in, 64–69
  type of, in header, 46–47
copyright protection, 196–198
  MM7 interface and, 173–174
Cyberlab, 190
Czech Republic, 183

data calling tone (DCT) in, 95
databases and MMS, 157–160
dates, 44
delivery context library (DELI), 102
delivery reports, 45, 64
delivery time setting, 45
developers' special handsets, 125
developing MMS applications, 15–16, 71–88
  ASCII files and, 83
  business case for MMS and, 184–185
  composing MMS application, 79–81, **80**
  construction of MMS and, 78–84
  content creation and, 86

developing MMS applications *(continued)*
  encoding MMS and, 83, **84**
  free downloadable tools for, 87, **87–88**
  handset support and, 81–82
  handsets for, 79–80
  HTTP and, 83
  media element editors in, 81–83, **82**
  MIME encoding and, 83
  multimedia editors in, 81
  portal software for, 80–81
  programs for, 85–87, **86–87**
  simulators for, 84–85
  SMIL and, 72–79
  SMIL players for, 84–85
  testing and, 124
  transforming MMS applications in, 81–83
  Web authoring tools and, 81
  wireless session protocol (WSP) and, 83
digital rights management (DRM), 18, 196–198
  MM7 interface and, 173–174
  transcoding and content adaptation in,, 67
digital signatures, transcoding and content adaptation in, 67
direct loading test MMS to handset, for testing, 124–125
distribution of MMS messages, 20
distribution paths for MMS, 189–192
DNA Finland, 205

EDGE connections, 125
editing of MMS messages (*See* transcoding and content adaptation)
email
  interface selection for, 24
  MM3 interface and, 149–151
  real time transport protocol (RTP) and, 151–153, **152**, **153**, 157
  session initiation protocol (SIP) and, 150
  simple notification and alarm protocol (SNAP) and, 150–151
  SMTP and, 150–151
eMelody, 112
encoding MMS, 83, **84**

end attribute, SMIL and, 77
enterprise MMS, 205–206
envelopes
MM4 envelopes in, 23
MM7 interface and, **166**
Ericsson, 73, 79–80, 185, 190, 199
Europe, 183
European Telecommunications Standards Institute (ETSI), 166, 199
example of MMS, 6–12
expiration date/time, 44
expiration times, 58
extensible hypertext markup language (XHTML), 95
extensible markup language (XML), 165

features of MMS vs. SMS, 12–13
fingerprinting, transcoding and content adaptation in, 67
Finland, 205, 206
Flash animation, 118
forwarding of MMS messages, 41, 64
free downloadable tools, 87, **87–88**
free running text content, 25
Full SMIL (*See also* SMIL), 72
future of MMS, 195–206
digital rights management and, 196–198
enterprise MMS, 205–206
Internet protocol multimedia systems (IMS) and, 203–204
roaming and, 200–201
sponsors and advertising in MMS, 204–205
standards and, 198–200
technical substitutes and, interaction with, 203–204
value added service providers (VASP) and, 202–203
value based pricing for MMS applications in, 201–202

gateways, WAP, 31, 209
general packet radio service (GPRS), 16, 31–32, 125, 199
BREW initiative, 191–192
GIF files, 15, 33, 114
animation and, 117–118
GISnet, 205
global system for mobile communication (GSM), 73
adaptive multirate (AMR) in, 108, 111
BREW initiative, 191–192
real time transport protocol (RTP) and, 152
GoldWave, 16, 65
GoLive (*See* Adobe GoLive)
graphic interchange format (*See* GIF files)
graphics, 114–118
animation in, 117–118
GIF files in, 114, 117–118
JPEG files in, 114
PNG files in, 114–115
size of, 115
standards for, 211
SVG files, 115–117, 203–204
synthetic video animation and, 117
VIM animation files in, 118
WBMP files in, 114–115
greeting cards (*See also* vCard), 118–119, 184
GSM SMIL, 73

Hallmark, 184
handsets, 4, 16
application development and, 81–82
application development and, 79–80
ASCII vs. binary MMS representation for, 126–127
composite capability/preference profiles (CC/PP) and, 96, 102
delivery context library (DELI) and, 102
developers' special handsets for, 125
direct loading test MMS to, 124–125
media formats and support and, 93, **93**, 96–102
m_notification_ind and m_retrieve_conf in, 127
multimedia messaging service center (MMSC) connection to, 21–23, **21**
simulators of (*See* simulators)
specifications for, 6

handsets *(continued)*
standards for, 198–200
support in, 122–123
user agent profile (UAProf) descriptions and, 96, 102, 225–259
headers
HTTP headers for MMS, 15–16, 38–40, **39**, 83, 209
MM1 Notification Request Envelope, 56–60
MM1 Notification Response Envelope, 59–60
MM1 Retrieval Request Envelope headers in, 60
MM1 Retrieval Response Envelope headers in, 61–63
MM1 Submit Response Envelope, 54–56
MM1_submit envelope and, 41–47, **42–43**
MM7 interface and, 165, 166–167, 174
real time transport protocol (RTP) and, 152, **152**
simple object access protocol (SOAP), 83, 165–170, **166**, 174
standards for, 209
Hello World application example, 36–37, **36, 37**
binary file of, for testing, 127–129, 131–136
delivery of message, 60–63
MM1 submission of, 54
high resolution media formats, 92–93, **93**
HSCSD connections, 125
HTML, special character sets and, 107, **107**
HTTP, 15–16, 30, 38–40, 83, 209
headers for MMS using, 15–16, 38–40, **39**, 83, 209
MM7 interface and, 165, 166–167, 174
hypertext transfer protocol (*See* HTTP)

image editors, 16
iMelody, 94, 112–113
income potential, in business case for MMS, 178, 179, **181–182**, 196
indexing, 159
Infrared Data Association (IrDA), 112
interfaces, 23–25, **24**
interfaces *(continued)*
MM1 (*See also* message structure and MM1), 23, 29–31, **29, 30**, 35–69, 164, 213–223
MM3 interface in (*See also* MM3 interface), 23, 24, 30, 147–161, **149**
MM4 envelopes in, 23
MM5 interface in, 23
MM6 interface in, 23
MM7 (*See also* MM7 interface), 23, 24, 30, 148, 163–175
MM9 interface in, 148
selection of, 24
International Association Naming Authority (IANA), 129
Internet Engineering Task Force (IETF), 207
Internet protocol (IP), 30
Internet protocol multimedia systems (IMS) and, 203–204
Internet servers, 3–4
interoperability, 13, 18, 198–200
Interoperability Group for MMS, 199
ISO 8859–1 special character set, 107, **107**
Italy, sale/use of mobile phones in, 2, 183

Japan, sale/use of mobile phones in, 2
Java (J2ME), 2
Job Dispatcher, 205
Joint Photographic Experts Group (*See* JPEG files), 114
Joulupukki TV, 205
JPEG files, 15, 33, 114, 212
Juniper Research, business case for MMS, 178, 179, 205

killer applications, 201

<layout> tags, 50–51
limitations of MMS vs. SMS, 12–13
Logica/CMG, 73, 185, 199
low–resolution media formats, 92–93, **93**

MacroMedia, 118
Macromedia Dreamweaver, 81

market predictions for MMS, 178
marketing potential in MMS, 185–186
media elements, 25–26
  editor tools for, 81–83, **82**
media formats, 89–120, **90–92**
  3GPP standard and, 90
  adaptive multirate (AMR) audio in, 94, 108
  <anchor> tag and, 104–106
  animation in, 117–118
  <area> tag and, 104, 106–107
  audio and voice codec selection in, 95, **95**
  audio files and, 108–114
  classification of, 92–93, **93**
  composite capability/preference profiles (CC/PP) and, 96, 102
  data calling tone (DCT) in, 95
  delivery context library (DELI) and, 102
  graphics and, 114–118
  handset support and, 93, **93**, 96–102
  iMelody in, 94, 112–113
  MIDI and SP–MIDI, 15, 94, 113–114, 212
  MMS Conformance Document and, 94
  resolution of, high vs. low, 92–93, **93**
  ringing tones text transfer language (RTTTL) format in, 94, 113
  selection of, 90
  SMAF in, 94
  SMIL and, 95, 96
  special character sets in, 107, **107**
  speech codecs and, 108
  standards for, 210
  testing, Web site SMS responder for, 102, **103–104**
  text files in, 104–107
  transcoding and, 95
  user agent profile (UAProf) handset descriptions and, 96, 102, 225–259
  vCard and vCalendar in, 118–119
Melody (*See also* iMelody), 112
message assembly, 26–28
message ID, 56
message passing, 28–30, **29**
message reference, 58
message store, 158–160, **159**
message structure and MM1, 35–69
  blocking messages in, 59
  content protection in, 63
  delivery of Hello World message in, 60–63
  expiration times in, 58
  forwarding of MMS messages in, 41
  Hello World application example in, 36–37, **36, 37**
  HTTP headers for MMS, 38–40, **39**
  <layout> tags in, 50–51
  message reference in, 58
  MIME encapsulation and, 47–49
  MM1_submit envelope and, 41–47, **42–43**
  model of operation and, 40–41
  notification of message arrival in, 56–60
  Notification Request Envelope, 56–60
  Notification Response Envelope, 59–60
  other MM1 envelopes for, 63–64, **64**
  <par> tags in, 51
  push messages in, 56
  request and response envelopes in, 37–40
  response of MMSC in, 54–56
  Retrieval Request Envelope headers in, 60
  Retrieval Response Envelope headers in, 61–63
  size of MMS messages in, 40–41
  SMIL 2 MIME encapsulation in, 51–53
  SMIL program in, 49–51
  status codes for, 55
  store and forward functionality in, 40–41
  submission of Hello World message in, 54
  transcoding and content adaptation in, 64–69
  wireless session protocol (WSP) in, 51
message types, 43
MIDI and SP–MIDI, 15, 94, 113–114, 212
MIME, 15, 27–29, **28, 29**, 81, 83, 209
  Boston Bruins sports highlights example in, 156
  email and, 149
  encapsulation of, in MM1 envelope, 47–49
  message store and, 160
  MM1 interface and, 47–49

MIME *(continued)*
real time transport protocol (RTP) and, 151–153, **152**, **153**, 157
SMIL conversion to, 51–53
SMIL templates and automatic MMS production in, 153–157, **155**
standards for, 210
wireless markup language (WML) MMS notification page and, 130–131
MIME::Lite, 51
MM1 interface (*See also* message structure and MMI), 23, 29–31, **29, 30**, 35–69, 81
blocking messages in, 59
complete MMS message sample on, 164, 213–223
content protection in, 63
delivery of Hello World message in, 60–63
forwarding of MMS messages in, 41
HTTP headers for MMS, 38–40, **39**
MIME encapsulation and, 47–49
MM1_submit envelope and, 41–47, **42–43**
model of operation and, 40–41
notification of message arrival in, 56–60
Notification Request Envelope, 56–60
Notification Response Envelope, 59–60
other envelopes for, 63–64, **64**
request and response envelopes on, 37–40
Retrieval Request Envelope headers in, 60
Retrieval Response Envelope headers in, 61–63
size of MMS messages in, 40–41
SMIL program in, 49–51
status codes for, 55
submission of Hello World message in, 54
wireless session protocol (WSP) in, 51
MM1_acknowledgment, 64
MM1_delivery_report, 64
MM1_forward, 64
MM1_read_reply_originator, 64
MM1_read_reply_recipient, 64
MM1_submit envelope, 41–47, **42–43**
MM3 interface, 23, 24, 30, 147–161, **149**
Boston Bruins sports highlights example in, 156
MM3 interface *(continued)*
databases and MMS in, 157–160
email and, 149–151
indexing in, 159
message store in, 158–160, **159**
real time transport protocol (RTP) and, 151–153, **152, 153**, 157
SMIL templates and automatic MMS production in, 153–157, **155**
MM4 envelopes, 23
MM5 interface in, 23
MM6 interface, 23
MM7 interface, 23, 24, 30, 148, 163–175
copyright protection and digital rights management in, 173–174
envelopes in, **166**
HTTP headers and, 165, 166–167, 174
messages in, 170–173, **170**
security and, 174
simple mail transfer protocol (SMTP) and, 165
simple object access protocol (SOAP) headers in, 165–170, **166**, 174
submit fields for, **171–172**
submit message on, 165–170, **166**
value added service provider (VASP) and, 163–175
VASP specific fields for, 171–173, **171–172**
MM9 interface in, 148
MMS Conformance Document, 25–26, 73, 199–200
media formats and, 94
size of message and, 104–105
MMS Conformance SMIL, 73, 76–79, **76**
MMS Conformance Specification, 73
m_notification_ind, 127
mobile phone use, 2
mobile photo service, 6–9, **7**, 190–191
model of operation, 32–33, **32**
MM1 interface and, 40–41
Motorola, 73, 199
MPEG files, 111
m_retrieve_conf, 127
multimedia editors, 81

multimedia Internet mail extension (*See* MIME)
multimedia messaging service center (MMSC), 17, 21, 79
computer connection to, 23–25, **21**
handset connection to, 21–23, **21**
interfaces to, 23–25, **24**
MM1 interface in (*See also* message structure and MM1; MM1 interface), 23, 29–31, **29, 30**, 35–69, 81, 164, 213–223
MM1 request/response envelopes and, 37–40
MM3 interface in (*See also* MM3 interface), 23, 24, 30, 147–161, **149**
MM4 envelopes in, 23
MM5 interface in, 23
MM6 interface in, 23
MM7 interface in (*See also* MM7 interface), 23, 24, 30, 148, 163–175
MM9 interface in, 148
response of, in MM1, 54–56
transcoding and content adaptation in, 64–69
wireless markup language (WML) notification page for, 131
wireless session protocol (WSP) and, 29, 31
multimedia players, desktop, 16

network nodes, 31, **31**
next steps in MMS (*See* future of MMS)
NMSS SDK API, 87
Nokia, 2, 73, 87, 185, 199
Nokia 5190, 16
Nokia Composer, 113
Nokia Developer's Suite, 80
Nokia Multimedia Converter, 109
Nokring format, 113
Norway, 183
notification of message arrival, MM1 interface and, 56–60
notification
m_notification_ind and m_retrieve_conf in, 127
push messages for, 129, 143–145
notification *(continued)*
short messaging service (SMS) message for, 129, 136–143
wireless markup language (WML) page for, 129, 130–131

Open Mobile Alliance (OMA), 129, 174, 207
Openwave, 185
opportunities and threats in MMS business, **181–182**
Ovum, business case for MMS and, 183

Paint Shop Pro, 16
<par> tags, 51
passing messages (*See* message passing)
Perl modules, MM1 interface and, 51
personal digital assistants (PDAs), 22
Philippines, 184
Phone.com, 185
photographs, mobile service for, 6–9, **7**, 190–191
Photoshop, 65
picture content, 25
platforms for mobile applications, 3–6, **4**
playback devices, 15
players, SMIL, 84–85
PNG files, 114–115
polyphonic ringtones, 203–204
portable network graphics (*See* PNG files)
portal software, 180
application development and, 80–81
testing and, 123–124
pricing fundamentals for MMS applications, 201–202
pricing vs. end user value, 186–189, **188–189**
value based pricing for MMS applications in, 201–202
priority setting for messages, 45
professional sports highlights service, 10–12, 156
programs, application development and, 85–87, **86–87**
protocols, 30–32

push messages, 56, 127
notification message using, 129, 143–145

QCELP, 111
Qualcomm, 185, 192

read reply setting, 46
read, reply, originator, 64
read, reply, recipient, 64
real time transport protocol (RTP),, 151–153, **152**, **153**, 157
receiver, in architecture of MMS, 20
regular pulse excited linear predictive code (RPE–LPC), 111
relay/server (proxy–relay), 20–21
reply charge, 45, 46
reply deadline, 45
request envelopes, MM1 interface and, 37–40
request status, 55
resolution of media formats, high vs. low, 92–93, **93**
resources for application development, 16–17
response envelopes, MM1 interface and, 37–40
retrieving messages, m–notification–ind and m–retrieve–conf in, 127
ringing tones text transfer language (RTTTL) format in, 94, 113
ringtones, 113, 203–204
roaming, 200–201
router, in architecture of MMS, 20

sales of mobile phones, 2
scalable vector graphics (*See* SVG files)
security
MM7 interface and, 174
transcoding and content adaptation in, 67
sender,, in architecture of MMS, 20
sender visibility, 46
servers
EDGE connection to, 125
GPRS connection to, 125
HSCSD connection to, 125
sending MMS from, for testing, 125
session description protocol (SDP), 153
session initiation protocol (SIP), 150
short messaging service (SMS), 3, 7, 12, 40, 56, 86–87, 148, 190
features and limitations of, vs. MMS, 12–13
MMS advantages over, 14
notification message using, 129, 136–143
short messaging service center (SMSC) in, 20, 40, 136
short messaging service center (SMSC), 20, 40, 136
Siemens, 73, 199
signaling in MMS, 32–33, **32**
simple mail transfer protocol (SMTP), 150–151, 165
simple notification and alarm protocol (SNAP), 150–151
simple object access protocol (SOAP) headers, 15, 83, 165–170, **166**, 174
simulators, 16, 84–85
Singapore, 184
Singapore MSS initiative, 189–191
Singtel, 189–191
size of applications/content for MMS, 15, 20, 32–33, 104–105
audio files and, 111, **111**
graphics and, 115
MM1 interface and, 40–41
SMIL and, 78
transcoding and content adaptation in, 66–69
SMAF, 94
Smart Telecom, 184
SMIL, 15, 26–28, **27**, 72–79, 210
<anchor> tag and, 104–106
<area> tag and, 104, 106–107
3GPP standard and, 72–73
begin and end attributes in, 77
Boston Bruins sports highlights example in, 156
code division multiple access (CDMA) and, 73
construction of MMS and, 78–84

SMIL *(continued)*
Full SMIL vs., 72
indexing in, 159
media formats and, 95, 96
message store and, 159–160
MIME encapsulation/conversion and, 51–53
MM1 interface and, 49–51, 49
MMS Conformance, 73, 76–79, **76**
multimedia editors and, 81
players for, 84–85
restrictions on messages and, 77–78, **78**
size of message and, 78
standards for, 212
synchronizing messages using, 72, **72**
tags in, **74–76**, 77, 78
templates in, automatic MMS production using, 153–157, **155**
testing and, 122
text files and, 104–105
user agent profile (UAProf) handset descriptions and, 102
versions/flavors of, 73–74, **74**
Web authoring tools and, 81
Sony Ericsson, 73, 199
Sony Ericsson T68i handset, 16, 22
special text character sets, 107, **107**
specifications (*See also* standards), 6, 207–212
MMS Conformance Specification, 73
speech codecs, 108, 212
sponsors and advertising in MMS, 204–205
sports highlights services, 10–12, 156
SPOT–XDE Pro, 109
standards, 5–6, 198–200, 207–212
3GPP standard in, 20
3GPP vs. 3GPP3, 200–201
business case for MMS and, 178
future of MMS and, 198–200
handsets, 198–200
media format, 210
MMS Conformance Document, 25–26
MMS Conformance Specification, 73
platforms for mobile applications vs., 5
roaming and, 200–201
status codes, MM1 interface and, 55
storage services, 180
store and forward functionality, 40–41, 158–160, **159**
store, message store, 158–160, **159**
strategic view of value chain participants, 180–182
streaming data, RTP, 151–153, **152**, **153**, 157
structured query language (SQL) (*See also* databases and MMS), 158
structured text content, 25
subject line, 46
submit fields, MM7 interface and,, 165–170, **166, 171–172**
subscriber identity module (SIM), 4
support for media elements
transcoding and content adaptation in, 67–69
SVG files, 115–117, 203–204, 211
animation and, 117
Sweden, sale/use of mobile phones in, 2
Swisscom, 183
Switzerland, 183
synchronized multimedia integration language (*See* SMIL)
synchronizing messages using SMIL, 72, **72**
synthetic audio, 111
synthetic video animation, 117

tags, SMIL, **74–76**, 77, 78
technical substitutes and MMS, interaction with, 203–204
Telenor, 183
templates, SMIL, automatic MMS production using, 153–157, **155**
testing, 17, 121–145
alpha testing in, 122
application developers programs and, 124
ASCII versus binary representation of MMS messages in, 126–127
beta testing in, 122
developers' special handsets for, 125
direct loading to handset in, 124–125
handset support and, 122–123

testing *(continued)*
Hello World binary file for, 127–129, 131–136
methods for, 123
portals for, 123–124
push message as notification for, 129, 143–145
sending MMS from own server, 125
SMIL and, 122
SMS message as notification in, 129, 136–143
Web–based MMS composers and, 123–124
wireless markup language (WML) notification in, 129, 130–131
XML editors and, 122
text content, 25
text files, 104–107
<anchor> tag and, 104–106
<area> tag and, 104, 106–107
size of message and, 104–105
SMIL and, 104–105
special character sets in, 107, **107**
vCard and vCalendar in, 118–119
wireless application protocol (WAP) and, 105
TIM, 183
timestamps, 44
real time transport protocol (RTP) and, 151–153, **152**
T–Mobile, 204–205
Toast, 111
transaction IDs, 43
transcoding, 64–69
digital rights management and, 196–198
media formats and, 95
transforming MMS applications, 81–83
transmission control protocol (TCP), 30
transport protocol data unit (TPDU), 136
TRUE BREW (*See* BREW initiative)

Unicode, standards for, 210
United States, sale/use of mobile phones in, 2
upgrades, 13
user agent profile (UAProf) handset descriptions, 96, 102, 225–259
user agents, in architecture of MMS, 22
UTF–8 standards, 211

value added service provider (VASP) (*See also* MM7 interface), 18, 163–175
copyright protection and digital rights management in, 173–174, 173, 202–203
MM7 interface in, 24
transcoding and content adaptation in, 66–69
value based pricing for MMS applications, 201–202
value chain in MSS, 179–189, **180**
vCalendar, 6, 118–119
vCard, 6, 118–119
version numbers, 3GPP, 43
video coding standards, 211
VIM animation files, 118
Vimatix, 118
Virtual Mechanics, 116
visibility of sender, 46
Visual Studio, 16
voice codec selection, 95
VoiceAge, 109

walled garden, 13, 16
WAP gateways, 31, 209
WAP–209–MMS Encapsulation, 129
WAV files, 15, 111
WBMP files, 114–115
Web authoring tools, 81
Web Dwarf, 116
Web–based MMS composers and, 123–124
Windows, 2
wireless application protocol (WAP), 12, 13–17, 56, 85, 178, 190, 199
binary file encoding and, 129
message store and, 160
push messages in, 127, 129, 143–145
standards for, 209
text files and, 105
wireless bitmap (*See* WBMP files)
Wireless Interim Naming Authority (WINA), 129

wireless markup language (WML),
notification page for MMS using, 129, 130–131
wireless session protocol (WSP), 29, 31, 81, 83
MM1 interface and, 51
World Wide Web Consortium (W3C), 96

XML editors, 122
XML::Twig, 51
X-Mms-3GPP-MMS-Version, 43
X-Mms-Content-Type, 46-47
X-Mms-Date-And-Time, 44
X-Mms-Delivery-Report, 45
X-Mms-Earliest-Delivery-Time, 45
X-Mms-Message-Id, 56
X-Mms-Message-Type, 43
X-Mms-Priority, 45
X-Mms-Read-Reply, 46
X-Mms-Recipient-Address, 43-44
X-Mms-Reply-Charging, 45
X-Mms-Reply-Charging-Id, 46
X-Mms-Reply-Charging-Size, 45
X-Mms-Reply-Deadline, 45
X-Mms-Request-Status, 55
X-Mms-Request-Status-Text in, 55
X-Mms-Sender-Address, 44
X-Mms-Sender-Visibility, 46
X-Mms-Subject, 46
X-Mms-Time-of-Expiry, 44
X-Mms-Transaction-Id, 43

ZOOMON, 117

ABOUT THE AUTHORS

Scott B. Guthery is CTO of Mobile-Mind, a wireless applications and software development firm headquartered in Boston, Massachusetts. He is also co-author of the popular *Mobile Application Development with SMS and the SIM Toolkit* (published by McGraw-Hill) and the definitive *Smart Cards: The Developer's Toolkit*. An active contributor to international wireless standards groups, Mr. Guthery chairs the Architecture Working Group of ETSI's Smart Card Platform Project. Prior to founding Mobile-Mind, he led the development of the first Java Card™ at Schlumberger and the Smart Card for Windows operating system at Microsoft.

Mary J. Cronin is President of Mobile-Mind and co-author of *Mobile Application Development with SMS and the SIM Toolkit*. She is also the author of numerous books and articles on business on the Internet, e-commerce, and mobile commerce, including *Doing Business on the Internet* and *Unchained Value*. Dr. Cronin serves on the Engineering Advisory Board of the Aurora Funds, a venture firm in North Carolina.